NYSTCE
006

CST
Biology
Teacher Certification Exam

By: Sharon Wynne, M.S.

XAMonline, INC.
Boston

XAMonline, Inc.
25 First Street, Suite 106
Cambridge, MA 02141
Toll Free: 1-800-509-4128
Email: info@xamonline.com
Web: www.xamonline.com
Fax: 1-617-583-5552

Library of Congress Cataloging-in-Publication Data

Wynne, Sharon A.
    CST Biology 006: Teacher Certification / Sharon A.
    Wynne. -2nd ed. ISBN 978-1-58197-289-4
    1. CST Biology 006.   2. Study Guides.   3. NYSTCE
    4. Teachers' Certification & Licensure.   5. Careers

Disclaimer:
The material presented in this publication is the sole work of XAMonline and was created independently from the National Education Association, Educational Testing Service, or any State Department of Education, National Evaluation Systems or other testing affiliates.

Between the time of publication and printing, state specific standards, testing formats, and website information may change. XAMonline developed the sample test questions and they reflect similar content as on real tests; however, they are not former tests. XAMonline assembles content that aligns with state standards, but makes no claims nor guarantees regarding test performance. Numerical scores are determined by testing companies such as NES or ETS and then are compared with individual state standards. A passing score varies from state to state.

Printed in the United States of America   œ - 1, 07

NYSTCE: CST Biology 006
ISBN: 978-1-58197-289-4

Table of Contents

SUBAREA I.   FOUNDATIONS OF SCIENTIFIC INQUIRY

COMPETENCY 1.0   UNDERSTAND THE RELATIONSHIPS AND COMMON THEMES THAT CONNECT MATHEMATICS, SCIENCE, AND TECHNOLOGY...........1

Skill 1.1   Analyzing similarities among systems in math, science, and technology (e.g., stability, equilibrium)................................1

Skill 1.2   Applying concepts and theories from mathematics and other sciences to a biological system 1

Skill 1.3   Analyzing the use of biology and other sciences in the design of a technological solution to a given problem................................1

Skill 1.4   Using the Internet, a variety of software (e.g., spreadsheets, graphing utilities, statistical packages, simulations), and technologies (e.g., graphing calculators, computers) to model and solve problems in mathematics, science, and technology 3

COMPETENCY 2.0   UNDERSTAND THE HISTORICAL AND CONTEMPORARY CONTEXTS OF BIOLOGICAL STUDY AND THE APPLICATIONS OF BIOLOGY AND BIOTECHNOLOGY TO SOCIETY 5

Skill 2.1   Recognizing the significance of key events in the history of biological study (e.g., development of the microscope, understanding the structure of DNA, use of animals in research, genomic research)................................5

Skill 2.2   Recognizing the contributions of diverse cultures and individuals to biological study 6

Skill 2.3   Evaluating the impact of social factors on biological study (e.g., restrictions on the development of human cloning techniques, demand for genetically modified agricultural crops, bioethics) 7

Skill 2.4   Interpreting the implications for society of recent developments in biology and biotechnology (e.g., medical technology, genetic engineering, wastewater treatment, food safety) 8

COMPETENCY 3.0   UNDERSTAND THE PROCESS OF SCIENTIFIC INQUIRY AND THE ROLE OF OBSERVATION, EXPERIMENTATION, AND COMMUNICATION IN EXPLAINING NATURAL PHENOMENA..........................10

Skill 3.1   Analyzing processes by which new scientific knowledge and hypotheses are generated...10

Skill 3.2   Analyzing ethical issues related to the process of scientific research and reporting 11

Skill 3.3   Evaluating the appropriateness of a specified experimental design to test a hypothesis...12

Skill 3.4   Demonstrating an ability to design a hypothesis-testing inquiry experiment...12

**COMPETENCY 4.0   UNDERSTAND THE PROCESSES OF GATHERING, ORGANIZING, REPORTING, AND INTERPRETING SCIENTIFIC DATA, AND APPLY THIS UNDERSTANDING IN THE CONTEXT OF BIOLOGICAL INVESTIGATIONS. **..................14

Skill 4.1   Evaluating the appropriateness of a given method or procedure for collecting data for a specified purpose 14

Skill 4.2   Selecting an appropriate and effective graphic representation (e.g., graph, table, diagram) for organizing, reporting, and analyzing given experimental data 15

Skill 4.3   Demonstrating the ability to appropriately set up and label graphs with dependent and independent variables 19

Skill 4.4   Applying procedures and criteria for reporting experimental protocols and data (e.g., use of statistical tests)..........................19

Skill 4.5   Analyzing relationships between factors (e.g., linear, exponential) as indicated by experimental data...........................20

**COMPETENCY 5.0   UNDERSTAND AND APPLY PRINCIPLES AND PROCEDURES OF MEASUREMENT USED IN THE BIOLOGICAL SCIENCES. **...22

Skill 5.1   Applying methods of measuring microscopic organisms and structures...22

Skill 5.2   Evaluating the appropriateness and limitations of units of measurement, measuring devices, or methods of measurement for given situations..23

Skill 5.3   Applying methods of measuring microscopic organisms and structures...23

COMPETENCY 6.0   UNDERSTAND THE USE OF EQUIPMENT, MATERIALS, CHEMICALS, AND LIVING ORGANISMS IN BIOLOGICAL STUDIES AND THE APPLICATION OF PROCEDURES FOR THEIR PROPER, SAFE, AND LEGAL USE. 24

Skill 6.1   Demonstrating knowledge of the appropriate use of given laboratory materials, instruments, and equipment (e.g., indicators, microscopes, centrifuges, spectrophotometers, chromatography equipment)..24

Skill 6.2   Applying proper methods for storing, identifying, dispensing, and disposing of chemicals and biological materials 25

Skill 6.3   Identifying sources of and interpreting information (e.g., material safety data sheets) regarding the proper, safe, and legal use of equipment, materials, and chemicals 26

Skill 6.4   Interpreting guidelines and regulations for the proper and humane procurement and treatment of living organisms in biological studies 26

Skill 6.5   Applying proper procedures for promoting laboratory safety (e.g., the use of safety goggles, universal health precautions) and responding to accidents and injuries in the biology laboratory 27

SUBAREA II.   CELL BIOLOGY AND BIOCHEMISTRY

COMPETENCY 7.0   UNDERSTAND CELL STRUCTURE AND FUNCTION, THE DYNAMIC NATURE OF CELLS, AND THE UNIQUENESS OF DIFFERENT TYPES OF CELLS. 29

Skill 7.1   Comparing and contrasting the cellular structures and functions of archaea, prokaryotes, and eukaryotes 29

Skill 7.2   Analyzing the primary functions, processes, products, and interactions of various cellular structures (e.g., lysosomes, microtubules, cell membrane) 30

Skill 7.3   Analyzing the importance of active and passive transport processes in maintaining homeostasis in cells and the relationships between these processes and the cellular membranes 33

COMPETENCY 8.0   UNDERSTAND CHEMISTRY AND BIOCHEMISTRY TO ANALYZE THE ROLE OF BIOLOGICALLY IMPORTANT ELEMENTS AND COMPOUNDS IN LIVING ORGANISMS. 36

Skill 8.1   Comparing and contrasting hydrogen, ionic, and covalent bonds and the conditions under which these bonds form and break apart 36

Skill 8.2   Relating the structure and function of carbohydrates, lipids, proteins (e.g., level of structure), nucleic acids, and inorganic compounds to cellular activities 37

Skill 8.3   Analyzing the properties of water and the significance of these properties to living organisms 40

Skill 8.4   Demonstrating an understanding of pH chemistry in biological systems 41

Skill 8.5   Analyze the structure and function of enzymes and factors that affect the rate of reactions 41

COMPETENCY 9.0   UNDERSTAND THE RAW MATERIALS, PRODUCTS, AND SIGNIFICANCE OF PHOTOSYNTHESIS AND CELLULAR RESPIRATION AND THE RELATIONSHIPS OF THESE PROCESSES TO CELL STRUCTURE AND FUNCTION. 43

Skill 9.1   Recognizing the significance of photosynthesis and respiration to living organisms 43

Skill 9.2   Identifying the overall chemical equations for the processes of respiration and photosynthesis 43

Skill 9.3   Demonstrating an understanding of ATP production through chemiosmosis in both photosynthesis and respiration 49

Skill 9.4   Analyzing factors that affect photosynthesis and respiration 49

Skill 9.5   Compare aerobic and anaerobic respiration 50

Skill 9.6   Evaluating the significance of chloroplast structure in photosynthesis and mitochondrion structure in respiration............51

COMPETENCY 10.0 UNDERSTAND THE STRUCTURE AND FUNCTION OF DNA AND RNA...53

Skill 10.1   Demonstrating an understanding of the mechanism of DNA replication, potential errors, and implications of these errors.........53

Skill 10.2   Analyzing the roles of DNA and ribosomal, messenger, and transfer RNA in protein synthesis 54

Skill 10.3   Analyzing the implications of mutations in DNA molecules for protein structure and function (e.g., sickle-cell anemia, cystic fibrosis)...56

Skill 10.4   Analyzing the control of gene expression in cells (e.g., lac operon in *E. coli*)...57

COMPETENCY 11.0 UNDERSTAND THE PROCEDURES INVOLVED IN THE ISOLATION, MANIPULATION, AND EXPRESSION OF GENETIC MATERIAL AND THE APPLICATION OF GENETIC ENGINEERING IN BASIC AND APPLIED RESEARCH. 59

Skill 11.1   Analyzing the role and applications of genetic engineering in the basic discoveries of molecular genetics (e.g., in medicine, agriculture) 59

Skill 11.2   Demonstrating an understanding of genetic engineering techniques (e.g., restriction enzymes, PCR, gel electrophoresis)..60

Skill 11.3   Analyzing the role of genetic engineering in the development of microbial cultures capable of producing valuable products (e.g., human insulin, growth hormone) 61

Skill 11.4   Recognizing the role of gene cloning in deriving nucleotide and amino acid sequences and the role of cloned genes as probes in determining the structure of more complex DNA molecules 61

Skill 11.5   Recognizing how genetic engineers design new biological products unavailable from natural sources and alter gene products by site-directed mutagenesis (e.g., transgenic plants and animals)..62

Skill 11.6 Recognizing the ethical, legal, and social implications of genetic engineering 65

COMPETENCY 12.0 UNDERSTAND THE CELL CYCLE, THE STAGES AND END PRODUCTS OF MEIOSIS AND MITOSIS, AND THE ROLE OF CELL DIVISION IN UNICELLULAR AND MULTICELLULAR ORGANISMS. 66

Skill 12.1 Describing general events in the cell cycle and analyzing the significance of these events 66

Skill 12.2 Interpreting the results of experiments relating to the eukaryotic cell cycle (e.g., cloning, polyploidy, tissue cultures, pharming) 67

Skill 12.3 Comparing chromosomal changes during the stages of meiosis and mitosis 71

Skill 12.4 Analyzing the significance of meiosis and fertilization in relation to the genetic diversity and evolution of multicellular organisms 73

Skill 12.5 Recognizing the relationship between an unrestricted cell cycle and cancer..73

Skill 12.6 Demonstrating an understanding of the process of cell differentiation, including the role of stem cells 74

SUBAREA III.   GENETICS AND EVOLUTION

COMPETENCY 13.0 UNDERSTAND CONCEPTS, PRINCIPLES, AND APPLICATIONS OF CLASSICAL AND MOLECULAR GENETICS. 76

Skill 13.1 Demonstrating an understanding of basic principles of heredity (e.g., dominance, codominance, incomplete dominance, segregation, independent assortment) 75

Skill 13.2 Analyzing techniques used to determine the presence of human genetic diseases (e.g., PKU, cystic fibrosis)..77

Skill 13.3 Analyzing genetic inheritance problems involving genotypic and phenotypic frequencies..79

Skill 13.4 Interpreting pedigree charts..79

Skill 13.5   Recognizing the role of nonnuclear inheritance (e.g., mitochondrial DNA) in phenotypic expression........................ 79

COMPETENCY 14.0 UNDERSTAND THE PRINCIPLES OF POPULATION GENETICS AND THE INTERACTION BETWEEN HEREDITY AND THE ENVIRONMENT, AND APPLY THIS KNOWLEDGE TO PROBLEMS INVOLVING POPULATIONS. 80

Skill 14.1   Evaluating conditions that affect allele frequency in a gene pool..80

Skill 14.2   Analyzing the relationship between an organism's phenotype for a particular trait and its selective advantage in a given environment (e.g., the human sickle-cell trait and malaria) 81

COMPETENCY 15.0 UNDERSTAND HYPOTHESES ABOUT THE ORIGINS OF LIFE, EVIDENCE SUPPORTING EVOLUTION, AND EVOLUTION AS A UNIFYING THEME IN BIOLOGY. 82

Skill 15.1   Evaluating evidence supporting various hypotheses about the origins of life 82

Skill 15.2   Analyzing the progression from simpler to more complex life forms (e.g., unicellular to colonial to multicellular) by various processes (e.g., endosymbiosis) 82

Skill 15.3   Assessing the significance of geological and fossil records in determining evolutionary histories and relationships of given organisms..84

Skill 15.4   Evaluating observations made in various areas of biology (e.g., embryology, biochemistry, anatomy) in terms of evolution 86

COMPETENCY 16.0 UNDERSTAND THE MECHANISMS OF EVOLUTION. 88

Skill 16.1   Recognizing sources of variation in a population 88

Skill 16.2   Analyzing relationships between changes in allele frequencies and evolution 88

Skill 16.3   Analyzing the implications of natural selection versus the inheritance of acquired traits in given situations 88

Skill 16.4   Comparing alternative mechanisms of evolution (e.g., gradualism, punctuated equilibrium)................................88

Skill 16.5   Analyzing factors that lead to speciation (e.g., geographic and reproductive isolation, genetic drift)................................89

COMPETENCY 17.0 UNDERSTAND THE PRINCIPLES OF TAXONOMY AND THE RELATIONSHIP BETWEEN TAXONOMY AND THE HISTORY OF EVOLUTION. 91

Skill 17.1   Analyzing criteria used to classify organisms (e.g., morphology, biochemical comparisons) 91

Skill 17.2   Interpreting a given phylogenetic tree or cladogram of related species 92

Skill 17.3   Demonstrating the ability to design and use taxonomic keys (e.g., dichotomous keys) 92

Skill 17.4   Demonstrating the ability to design and use taxonomic keys (e.g., dichotomous keys) 93

Skill 17.5   Relating changes in the structure and organization of the classification system to developments in biological thought (e.g., evolution, modern genetics) 94

SUBAREA IV.   BIOLOGICAL UNITY AND DIVERSITY AND LIFE PROCESSES

COMPETENCY 18.0 UNDERSTAND THE UNITY AND DIVERSITY OF LIFE, INCLUDING COMMON STRUCTURES AND FUNCTIONS. 95

Skill 18.1   Analyzing characteristics of living organisms (e.g., differences between living organisms and nonliving things) 95

Skill 18.2   Recognizing levels of organization (e.g., cells, tissues, organs)....95

Skill 18.3   Comparing and analyzing the basic life functions carried out by living organisms (e.g., obtaining nutrients, excretion, reproduction) 96

Skill 18.4   Recognizing the role of physiological processes (e.g., active transport) that contribute to homeostasis and dynamic equilibrium 97

Skill 18.5   Recognizing the relationship of structure and function in all living things................................98

COMPETENCY 19.0 UNDERSTAND THE GENERAL CHARACTERISTICS, FUNCTIONS, AND ADAPTATIONS OF PRIONS, VIRUSES, BACTERIA, PROTOCTISTS (PROTISTS), AND FUNGI. 99

Skill 19.1   Comparing the structure and processes of prions and viruses to cells 99

Skill 19.2   Comparing archaebacteria and eubacteria................................99

Skill 19.3   Analyzing the processes of chromosome and plasmid replication and gene transfer in bacteria.................................100

Skill 19.4   Comparing the structure and function of protoctists (protists) 101

Skill 19.5   Recognizing the significance of prions, viruses, retroviruses, bacteria, protoctists (protists), and fungi in terms of their beneficial uses or deleterious effects...101

COMPETENCY 20.0 UNDERSTAND THE GENERAL CHARACTERISTICS, LIFE FUNCTIONS, AND ADAPTATIONS OF PLANTS.......................................103

Skill 20.1   Comparing structures and their functions in nonvascular and vascular plants (e.g., mosses, ferns, conifers) 103

Skill 20.2   Analyzing reproduction and development in the different divisions of plants 104

Skill 20.3   Analyzing the structures and forces involved in transport in plants...106

Skill 20.4   Evaluating the evolutionary and adaptive significance of plant structures (e.g., modified leaves, colorful flowers).......................107

COMPETENCY 21.0 UNDERSTAND THE GENERAL CHARACTERISTICS, LIFE FUNCTIONS, AND ADAPTATIONS OF ANIMALS. 108

Skill 21.1   Identifying general characteristics of the embryonic development of invertebrates and vertebrates 108

Skill 21.2   Comparing and contrasting the life cycles of invertebrates and vertebrates 109

Skill 21.3   Demonstrating an understanding of physiological processes (e.g., excretion, respiration, aging) of animals and their significance...110

Skill 21.4   Recognizing the relationship between structure and function in given animal species 110

Skill 21.5   Analyzing the adaptive and evolutionary significance of animal behaviors and structures 112

SUBAREA V.   HUMAN BIOLOGY

COMPETENCY 22.0   UNDERSTAND THE STRUCTURES AND FUNCTIONS OF THE HUMAN SKELETAL, MUSCULAR, AND INTEGUMENTARY SYSTEMS; COMMON MALFUNCTIONS OF THESE SYSTEMS; AND THEIR HOMEOSTATIC RELATIONSHIPS WITHIN THE BODY...113

Skill 22.1   Comparing the structures, locations, and functions of the three types of muscles 113

Skill 22.2   Demonstrating an understanding of the mechanism of skeletal muscle contraction 113

Skill 22.3   Demonstrating an understanding of the movements of body joints in terms of muscle and bone arrangement and action 114

Skill 22.4   Relating the structure of the skin to its functions 114

Skill 22.5   Demonstrating an understanding of possible causes, effects, prevention, and treatment of malfunctions of the skeletal, muscular, and integumentary systems (e.g., arthritis, skin cancer, scoliosis, osteoporosis) 115

COMPETENCY 23.0   UNDERSTAND THE STRUCTURES AND FUNCTIONS OF THE HUMAN RESPIRATORY AND EXCRETORY SYSTEMS, COMMON MALFUNCTIONS OF THESE SYSTEMS, AND THEIR HOMEOSTATIC RELATIONSHIPS WITHIN THE BODY. 117

Skill 23.1   Demonstrating an understanding of the relationship between surface area and volume and the role of that relationship in the function of the respiratory and excretory systems 117

Skill 23.2   Analyzing the mechanism of breathing and the process of gas exchange between the lungs and blood and between blood and tissues 117

Skill 23.3   Analyzing the role of the kidneys in osmoregulation and waste removal from the blood and the factors that influence nephron function...118

Skill 23.4   Demonstrating an understanding of possible causes, effects, prevention, and treatment of malfunctions of the respiratory and excretory systems (e.g., emphysema, nephritis) 119

COMPETENCY 24.0 UNDERSTAND THE STRUCTURES AND FUNCTIONS OF THE HUMAN CIRCULATORY AND IMMUNE SYSTEMS, COMMON MALFUNCTIONS OF THESE SYSTEMS, AND THEIR HOMEOSTATIC RELATIONSHIPS WITHIN THE BODY.........................120

Skill 24.1   Demonstrating an understanding of the structure, function, and regulation of the heart and the factors that influence cardiac output 120

Skill 24.2   Analyzing changes in the circulatory system (e.g., vessel structure and function) and their influence on blood composition and blood flow (e.g., blood cell diversity).................121

Skill 24.3   Demonstrating an understanding of the possible causes, effects, prevention, and treatment of malfunctions of the circulatory system (e.g., hypertension).......................................122

Skill 24.4   Demonstrating an understanding of the structure, function, and regulation of the immune system (e.g., cell-mediated and humoral responses)...123

Skill 24.5   Demonstrating an understanding of the possible causes, effects, prevention, and treatment of malfunctions of the immune system (e.g., autoimmune diseases, transplant rejection)..124

COMPETENCY 25.0 UNDERSTAND HUMAN NUTRITION AND THE STRUCTURES AND FUNCTIONS OF THE HUMAN DIGESTIVE SYSTEM AND ACCESSORY ORGANS, COMMON MALFUNCTIONS OF THE DIGESTIVE SYSTEM, AND ITS HOMEOSTATIC RELATIONSHIPS WITHIN THE BODY.........................126

Skill 25.1   Demonstrating an understanding of the roles in the body of the basic nutrients found in foods (e.g., carbohydrates, vitamins, water) 126

Skill 25.2   Demonstrating an understanding of the processes of mechanical and chemical digestion in the digestive system, including contributions of accessory organs 127

Skill 25.3   Recognizing the process by which nutrients are transported from inside the small intestine to other parts of the body 127

Skill 25.4   Demonstrating an understanding of the possible causes, effects, prevention, and treatment of common malfunctions of the digestive system (e.g., ulcers, appendicitis, eating disorders) 129

COMPETENCY 26.0 UNDERSTAND THE STRUCTURES AND FUNCTIONS OF THE HUMAN NERVOUS AND ENDOCRINE SYSTEMS, COMMON MALFUNCTIONS OF THESE SYSTEMS, AND THEIR HOMEOSTATIC RELATIONSHIPS WITHIN THE BODY. 130

Skill 26.1   Demonstrating an understanding of the structures and functions of the central and peripheral nervous systems 130

Skill 26.2   Demonstrating an understanding of the location and function of the major endocrine glands and the function of their associated hormones 131

Skill 26.3   Demonstrating an understanding of the transmission of nerve impulses within and between neurons and the influence of drugs and other chemicals on that transmission 132

Skill 26.4   Evaluating the role of feedback mechanisms in homeostasis (e.g., role of hormones, neurotransmitters) 133

Skill 26.5   Demonstrating an understanding of the possible causes, effects, prevention, and treatment of malfunctions of the nervous and endocrine systems (e.g., diabetes, brain disorders) 133

COMPETENCY 27.0 UNDERSTAND THE STRUCTURES AND FUNCTIONS OF THE HUMAN REPRODUCTIVE SYSTEMS, THE PROCESSES OF EMBRYONIC DEVELOPMENT, COMMON MALFUNCTIONS OF THE REPRODUCTIVE SYSTEMS, AND THEIR HOMEOSTATIC RELATIONSHIPS WITHIN THE BODY..**136**

Skill 27.1   Recognizing the role of hormones in controlling the development and functions of the male and female reproductive systems.. 135

Skill 27.2   Demonstrating an understanding of gametogenesis, fertilization, and birth control.. 135

Skill 27.3   Demonstrating an understanding of embryonic and fetal development and the potential effects of drugs, alcohol, and nutrition on this process.. 136

Skill 27.4   Demonstrating an understanding of the possible causes, effects, prevention, and treatment of malfunctions of the reproductive systems (e.g., infertility, birth defects)...................... 137

SUBAREA VI.   **ECOLOGY**

COMPETENCY 28   UNDERSTAND THE CHARACTERISTICS OF POPULATIONS AND COMMUNITIES AND USE THIS KNOWLEDGE TO INTERPRET POPULATION GROWTH AND INTERACTIONS OF ORGANISMS WITHIN AN ECOSYSTEM. **140**

Skill 28.1   Demonstrating an understanding of factors that affect population size and growth rate (e.g., carrying capacity, limiting factors) 140

Skill 28.2   Determining and interpreting population growth curves............... 140

Skill 28.3   Analyzing relationships among organisms in a community (e.g., competition, predation, symbiosis) 142

Skill 28.4   Evaluating the effects of population density on the environment.. 143

COMPETENCY 29.0 UNDERSTAND THE DEVELOPMENT AND
STRUCTURE OF ECOSYSTEMS AND THE
CHARACTERISTICS OF MAJOR BIOMES. 144

Skill 29.1 Demonstrating an understanding of the flow of energy through
the trophic levels of an ecosystem 144

Skill 29.2 Comparing the strengths and limitations of various pyramid
models (e.g., biomass, numbers, energy) 145

Skill 29.3 Identifying the characteristics and geographic distribution of
major biomes 146

Skill 29.4 Recognizing the effect of biome degradation and destruction
on biosphere stability (e.g., climate changes, deforestation,
reduction of species diversity) 151

COMPETENCY 30.0 UNDERSTAND THE CONNECTIONS WITHIN AND
AMONG THE BIOGEOCHEMICAL CYCLES AND
ANALYZE THEIR IMPLICATIONS FOR LIVING
THINGS. 153

Skill 30.1 Recognizing the importance of the processes involved in
material cycles (e.g., water, carbon, nitrogen, phosphorus) 153

Skill 30.2 Demonstrating an understanding of the role of decomposers in
nutrient cycling in ecosystems 153

Skill 30.3 Analyzing the role of respiration and photosynthesis in
biogeochemical cycling 154

Skill 30.4 Evaluating the effects of limiting factors on ecosystem
productivity (e.g., light intensity, gas concentrations, mineral
availability) 154

COMPETENCY 31.0 UNDERSTAND CONCEPTS OF HUMAN ECOLOGY
AND THE IMPACT OF HUMAN DECISIONS AND
ACTIVITIES ON THE PHYSICAL AND LIVING
ENVIRONMENT. 156

Skill 31.1 Recognizing the importance and implications of various factors
(e.g., nutrition, public health, geography, climate) for human
population dynamics 156

Skill 31.2 Predicting the impact of the human use of natural resources
(e.g., forests, rivers) on the stability of ecosystems 156

Skill 31.3   Recognizing the importance of maintaining biological diversity
(e.g., pharmacological products, stability of ecosystems)............157

Skill 31.4   Evaluating methods and technologies that reduce or mitigate
environmental degradation..158

Skill 31.5   Demonstrating an understanding of the concept of
stewardship and ways in which it is applied to the environment..159

SUBAREA VII. ....FOUNDATIONS OF SCIENTIFIC INQUIRY:
                                               CONSTRUCTED RESPONSE ASSIGNMENT

Sample Constructed Response Assignment...160

Sample Test...163

Answer Key 187

Rigor Table 188

Rationales with Sample Questions 189

Great Study and Testing Tips!

What to study in order to prepare for the subject assessments is the focus of this study guide, but equally important is *how* you study.

You can increase your chances of truly mastering the information by taking some simple, but effective steps.

Study Tips:

1. <u>Some foods aid the learning process</u>. Foods such as milk, nuts, seeds, rice, and oats help your study efforts by releasing natural memory enhancers called CCKs (*cholecystokinins*) composed of *tryptophan*, *choline*, and *phenylalanine*. All of these chemicals enhance the neurotransmitters associated with memory. Before studying, try a light, protein-rich meal of eggs, turkey, and fish. All of these foods release the memory enhancing chemicals. The better the connections, the more you comprehend.

Likewise, before you take a test, stick to a light snack of energy boosting and relaxing foods. A glass of milk, a piece of fruit, or some peanuts all release various memory-boosting chemicals and help you relax and focus on the subject at hand.

2. <u>Learn to take great notes</u>. A by-product of our modern culture is that we have grown accustomed to getting our information in short doses (e.g., TV news sound bites or USA Today style newspaper articles).

Consequently, we've subconsciously trained ourselves to assimilate information better in <u>neat little packages</u>. If your notes are scrawled all over the paper, it fragments the flow of the information. Strive for clarity. Newspapers use a standard format to achieve clarity. Your notes can be much clearer through use of proper formatting. A very effective format is called the *"Cornell Method."*

> Take a sheet of loose-leaf lined notebook paper and draw a line all the way down the paper about 1-2" from the left-hand edge.

> Draw another line across the width of the paper about 1-2" up from the bottom. Repeat this process on the reverse side of the page.

Look at the highly effective result. You have ample room for notes, a left hand margin for special emphasis items or inserting supplementary data from the textbook, a large area at the bottom for a brief summary, and a little rectangular space for just about anything you want.

3. <u>**Get the concept then the details**</u>. Too often we focus on the details and fail to gather an understanding of the concept. If you simply memorize dates, places, and names, you may well miss the whole point of the subject.

Putting concepts in your own words can increase your understanding. If you are working from a textbook, automatically summarize each paragraph in your mind. If you are outlining text, don't simply copy the author's words, *rephrase* them in your own words.

You remember your own thoughts and words much better than someone else's, and subconsciously tend to associate the important details to the core concepts.

4. <u>**Ask Why?**</u> Pull apart written material paragraph by paragraph and don't forget the captions under the illustrations.

Example: If the heading is "Stream Erosion", flip it around to read "Why do streams erode?" Then answer the question.

If you train your mind to think in a series of questions and answers, not only will you learn more, but you will decrease your test anxiety by increasing your familiarity with the question and answer process.

5. <u>**Read for reinforcement and future needs**</u>. Even if you only have ten minutes, put your notes or a book in your hand. Your mind is similar to a computer, you have to input data in order to process it. *By reading, you are creating the neural connections for future retrieval.* The more times you read something, the more you reinforce the learning of ideas.

Even if you don't fully understand something on the first pass, *your mind stores much of the material for later recall.*

6. <u>**Relax to learn so go into exile**</u>. Our bodies respond to an inner clock called biorhythms. Burning the midnight oil works well for some people, but not everyone.

If possible, set aside a particular place to study that is free of distractions. Shut off the television, cell phone, and pager and exile your friends and family during your study period.

If silence really bothers you, try background music. Light classical music at a low volume has been shown to aid in concentration. Music that evokes pleasant emotions without lyrics is highly suggested. Try just about anything by Mozart. It relaxes you.

7. Use arrows not highlighters. At best, it's difficult to read a page full of yellow, pink, blue, and green streaks. Try staring at a neon sign for a while and you'll soon see that the horde of colors obscure the message.

A quick note, a brief dash of color, an underline, and an arrow pointing to a particular passage is much clearer than a horde of highlighted words.

8. Budget your study time. Although you shouldn't ignore any of the material, *allocate your available study time in the same ratio that topics may appear on the test.*

Testing Tips:

1. **Get smart, play dumb. Don't read anything into the question.** Don't make an assumption that the test writer is looking for something else than what is asked.

2. **Read the question and all the choices _twice_ before answering the question.** You may miss something by not carefully reading and re-reading both the question and the answers.

If you really don't have a clue as to the right answer, leave it blank on the first time through. Go on to the other questions, as they may provide a clue as to how to answer the skipped questions.

If later on, you still can't answer the skipped ones . . . **_Guess._** The only penalty for guessing is that you _might_ get it wrong. Only one thing is certain; if you don't put anything down, you will get it wrong!

3. **Turn the question into a statement.** Look at the wording of the questions. The syntax of the question usually provides a clue. Does it seem more familiar as a statement rather than as a question? Does it sound strange?

By turning a question into a statement, you may be able to spot if an answer sounds right, and it may also trigger memories of material you have read.

4. **Look for hidden clues.** It's actually very difficult to compose multiple-foil (choice) questions without giving away part of the answer in the options presented.

In most multiple-choice questions you can often readily eliminate one or two of the potential answers. This leaves you with only two real possibilities and automatically your odds go to fifty-fifty with very little work.

5. **Trust your instincts.** On questions that you aren't really certain about, go with your basic instincts. **Your first impression on how to answer a question is usually correct.**

6. **Mark your answers directly on the test booklet.** Don't bother trying to fill in the optical scan sheet on the first pass through the test.

Just be very careful not to miss-mark your answers when you eventually transcribe them to the scan sheet.

7. **Watch the clock!** You have a set amount of time to answer the questions. Don't get bogged down trying to answer a single question at the expense of 10 questions you can more readily answer.

SUBAREA I.   FOUNDATIONS OF SCIENTIFIC INQUIRY

COMPETENCY 1.0   UNDERSTAND THE RELATIONSHIPS AND COMMON
                    THEMES THAT CONNECT MATHEMATICS, SCIENCE,
                    AND TECHNOLOGY

Skill 1.1   Analyzing similarities among systems in math, science, and
               technology (e.g., stability, equilibrium)

Math, science, and technology share many common themes. All three use
models, diagrams, and graphs to simplify a concept for analysis and
interpretation. Patterns observed in these systems lead to predictions based on
these observations. Another common theme among these three systems is
equilibrium. **Equilibrium** is a state in which forces are balanced, resulting in
stability. Static equilibrium is stability due to a lack of changes, and dynamic
equilibrium is stability due to a balance between opposing forces.

[handwritten margin note: 2 types of equilibrium — static: lack of changes — dynamic: balance between opposing forces.]

Skill 1.2   Applying concepts and theories from mathematics and other
               sciences to a biological system

The knowledge and use of basic mathematical concepts and skills is a necessary
aspect of scientific study. Science depends on data and the manipulation of data
requires knowledge of mathematics. Understanding of basic statistics, graphs,
charts, and algebra are of particular importance. Scientists must be able to
understand and apply the statistical concepts of mean, median, mode, and range
to sets of scientific data. In addition, scientists must be able to represent data
graphically and interpret graphs and tables. Finally, scientists often use basic
algebra to solve scientific problems and design experiments. For example, the
substitution of variables is a common strategy in experiment design. Also, the
ability to determine the equation of a curve is valuable in data manipulation,
experimentation, and prediction.

Skill 1.3   Analyzing the use of biology and other sciences in the design
               of a technological solution to a given problem

Science and technology are interdependent as advances in technology often
lead to new scientific discoveries and new scientific discoveries often lead to new
technologies. Scientists use technology to enhance the study of nature and
solve problems that nature presents. Technological design is the identification of
a problem and the application of scientific knowledge to solve the problem.
While technology and technological design can provide solutions to problems
faced by humans, technology must exist within nature and cannot contradict
physical or biological principles. In addition, technological solutions are
temporary and new technologies typically provide better solutions in the future.

Monetary costs, available materials, time, and available tools also limit the scope of technological design and solutions. Finally, technological solutions have intended benefits and unexpected consequences. Scientists must attempt to predict the unintended consequences and minimize any negative impact on nature or society.

The problems and needs, ranging from very simple to highly complex, that technological design can solve are nearly limitless. Disposal of toxic waste, routing of rainwater, crop irrigation, and energy creation are but a few examples of real-world problems that scientists address or attempt to address with technology.

The technological design process has five basic steps:

1. Identify a problem
2. Propose designs and choose between alternative solutions
3. Implement the proposed solution
4. Evaluate the solution and its consequences
5. Report results

After the identification of a problem, the scientist must propose several designs and choose between the alternatives. Scientists often utilize simulations and models in evaluating possible solutions.

Implementation of the chosen solution involves the use of various tools depending on the problem, solution, and technology. Scientists may use both physical tools and objects and computer software.

After implementation of the solution, scientists evaluate the success or failure of the solution against pre-determined criteria. In evaluating the solution, scientists must consider the negative consequences as well as the planned benefits.

Finally, scientists must communicate results in different ways – orally, written, models, diagrams, and demonstrations.

Example:

Problem – toxic waste disposal

Chosen solution – genetically engineered microorganisms to digest waste

Implementation – use genetic engineering technology to create organism capable of converting waste to environmentally safe product

Evaluate – introduce organisms to waste site and measure formation of products and decrease in waste; also evaluate any unintended effects

Report – prepare a written report of results complete with diagrams and figures

Identify a design problem and propose possible solutions, considering such constraints as tools, materials, time, costs, and laws of nature.

In addition to finding viable solutions to design problems, scientists must consider such constraints as tools, materials, time, costs, and laws of nature. Effective implementation of a solution requires adequate tools and materials. Scientists cannot apply scientific knowledge without sufficient technology and appropriate materials (e.g. construction materials, software). Technological design solutions always have costs. Scientists must consider monetary costs, time costs, and the unintended effects of possible solutions. Types of unintended consequences of technological design solutions include adverse environmental impact and safety risks. Finally, technology cannot contradict the laws of nature. Technological design solutions must work within the framework of the natural world.

In evaluating and choosing between potential solutions to a design problem, scientists utilize modeling, simulation, and experimentation techniques. Small-scale modeling and simulation help test the effectiveness and unexpected consequences of proposed solutions while limiting the initial costs. Modeling and simulation may also reveal potential problems that scientists can address prior to full-scale implementation of the solution. Experimentation allows for evaluation of proposed solutions in a controlled environment where scientists can manipulate and test specific variables.

Skill 1.4   **Using the Internet, a variety of software (e.g., spreadsheets, graphing utilities, statistical packages, simulations), and technologies (e.g., graphing calculators, computers) to model and solve problems in mathematics, science, and technology**

Biologists use a variety of tools and technologies to perform tests, collect and display data, and analyze relationships. Examples of commonly used tools include computer-linked probes, spreadsheets, and graphing calculators.

Biologists use computer-linked probes to measure various environmental factors including temperature, dissolved oxygen, pH, ionic concentration, and pressure. The advantage of computer-linked probes, as compared to more traditional observational tools, is that the probes automatically gather data and present it in an accessible format. This property of computer-linked probes eliminates the need for constant human observation and manipulation.

Biologists use spreadsheets to organize, analyze, and display data. For example, conservation ecologists use spreadsheets to model population growth and development, apply sampling techniques, and create statistical distributions to analyze relationships. Spreadsheet use simplifies data collection and manipulation and allows the presentation of data in a logical and understandable format.

Graphing calculators are another technology with many applications to biology. For example, biologists use algebraic functions to analyze growth, development and other natural processes.

Graphing calculators can manipulate algebraic data and create graphs for analysis and observation. In addition, biologists use the matrix function of graphing calculators to model problems in genetics. The use of graphing calculators simplifies the creation of graphical displays including histograms, scatter plots, and line graphs. Biologists can also transfer data and displays to computers for further analysis. Finally, biologists connect computer-linked probes, used to collect data, to graphing calculators to ease the collection, transmission, and analysis of data.

COMPETENCY 2.0   UNDERSTAND THE HISTORICAL AND
                  CONTEMPORARY CONTEXTS OF BIOLOGICAL
                  STUDY AND THE APPLICATIONS OF BIOLOGY AND
                  BIOTECHNOLOGY TO SOCIETY

Skill 2.1   Recognizing the significance of key events in the history of
             biological study (e.g., development of the microscope,
             understanding the structure of DNA, use of animals in
             research, genomic research)

Anton van Leeuwenhoek is known as the father of microscopy. In the 1650s,
Leeuwenhoek began making tiny lenses that produced magnifications up to 300x.
He was the first to see and describe bacteria, yeast plants, and the microscopic
life found in water.   Over the years, light microscopes have advanced to produce
greater clarity and magnification. The scanning electron microscope (SEM) was
developed in the 1950s. Instead of light, a beam of electrons passes through the
specimen. Scanning electron microscopes have a resolution about one
thousand times greater than light microscopes. The disadvantage of the SEM is
that the chemical and physical methods used to prepare the sample result in the
death of the specimen.

In the late 1800s, Pasteur discovered the role of microorganisms in the cause of
disease, pasteurization, and the rabies vaccine. Koch took this observation one
step further by formulating a theory that specific pathogens caused specific
diseases. Scientists still use **Koch's postulates** as guidelines in the field of
microbiology. The guidelines state that the same pathogen must be found in
every diseased person, the pathogen must be isolated and grown in culture, the
pathogen must induce disease in experimental animals, and the same pathogen
must be isolated from the experimental animal.

The discovery of the structure of DNA was another key event in biological study.
In the 1950s, James Watson and Francis Crick identified the structure of a DNA
molecule as that of a double helix. This structure made it possible to explain
DNA's ability to replicate and to control the synthesis of proteins.

The use of animals in biological research has expedited many scientific
discoveries. Animal research has allowed scientists to learn more about animal
biological systems, including the circulatory and reproductive systems. One
significant use of animals is for the testing of drugs, vaccines, and other products
(such as perfumes and shampoos) before use or consumption by humans.
The debate about the ethical treatment of animals has been ongoing since the
introduction of animals in research. Many people believe the use of animals in
research is cruel and unnecessary. Animal use is federally and locally regulated.
The purpose of the Institutional Animal Care and Use Committee (IACUC) is to
oversee and evaluate all aspects of an institution's animal care and use program.

Skill 2.2   Recognizing the contributions of diverse cultures and individuals to biological study

Curiosity is the heart of science. This is why so many diverse people are drawn to it. In the area of zoology one of the most recognized scientists is Jane Goodall. Goodall is known for her research with chimpanzees in Africa. She has spent many years abroad conducting long term studies of chimp interactions, and returns from Africa to lecture and provide information about Africa, the chimpanzees, and her institute located in Tanzania.

In the area of chemistry we recognize Dorothy Crowfoot Hodgkin. She studied at Oxford and won the Nobel Prize of Chemistry in 1964 for recognizing the shape of the vitamin B-12.

Florence Nightingale was a 19th century nurse who shaped the nursing profession. Nightingale was born into wealth and shocked her family by choosing to study health reforms for the poor in lieu of attending the expected social events. She studied nursing in Paris and became involved in the Crimean War. The British lacked supplies and the secretary of war asked for Nightingale's assistance. She earned her nickname walking the floors at night checking on patients and writing letters to British officials demanding supplies.

In 1903, the Nobel Prize in Physics was jointly awarded to three individuals: Marie Curie, Pierre Curie, and Becquerel. Marie Curie was the first woman ever to receive this prestigious award. In addition, she received the Nobel Prize in chemistry in 1911, making her the only person to receive two Nobel awards in science. Ironically, her cause of death in 1934 was of overexposure to radioactivity, the research for which she was so respected.

Neil Armstrong is an American icon. He will always be symbolically linked to our aeronautics program. An astronaut and naval aviator, he is best known for being the first human to set foot on the Moon.

Sir Alexander Fleming was a pharmacologist from Scotland who isolated the antibiotic penicillin from a fungus in 1928. Flemming also noted that bacteria developed resistance whenever too little penicillin was used or when it was used for too short a period, a key problem we still face today.

Skill 2.3 **Evaluating the impact of social factors on biological study (e.g., restrictions on the development of human cloning techniques, demand for genetically modified agricultural crops, bioethics)**

Society, as a whole, impacts biological research. The pressure from the majority of society has led to bans and restrictions on human cloning research. The United States government and the governments of many other countries have restricted human cloning. The U.S. legislature has banned the use of federal funds for the development of human cloning techniques. Some individual states have banned human cloning regardless of where the funds originate.

The demand for genetically modified crops by society and industry has steadily increased over the years. Genetic engineering in the agricultural field has led to improved crops for human use and consumption. Crops are genetically modified for increased growth and insect resistance because of the demand for larger and greater quantities of produce.

With advances in biotechnology come those in society who oppose it. Ethical questions come into play when discussing animal and human research. Does it need to be done? What are the effects on humans and animals? There are not absolute right or wrong answers to these questions. There are governmental agencies in place to regulate the use of humans and animals for research.

Science and technology are often referred to as a "double-edged sword". Although advances in medicine have greatly improved the quality and length of life, certain moral and ethical controversies have arisen. Unforeseen environmental problems may result from technological advances. Advances in science have led to an improved economy through biotechnology as applied to agriculture, yet it has put our health care system at risk and has caused the cost of medical care to skyrocket. Society depends on science, yet is necessary that the public be scientifically literate and informed in order to allow potentially unethical procedures to occur. Especially vulnerable are the areas of genetic research and fertility. It is important for science teachers to stay abreast of current research and to involve students in critical thinking and ethics whenever possible.

Skill 2.4 **Interpreting the implications for society of recent developments in biology and biotechnology (e.g., medical technology, genetic engineering, wastewater treatment, food safety)**

Medical Technology

The relationship between technology and society is synergistic. Applied research is of great value because it is directly useful to us; it deals with issues like AIDS, Tuberculosis, HPV, and Parkinson's disease. Technological advances have greatly increased the overall quality of medical care. Advances in medical technology benefit society in many ways including the expansion of medical research and the development of novel diagnostics and therapeutics.

Genetic Engineering

Research in molecular genetics is highly technical and very useful to society. Because of its use to humanity, molecular genetics receives generous funding from various sources and support from a large segment of the scientific community. Scientists use a number of techniques to isolate genes. The foremost technique is cDNA cloning, which is very reliable. The objective of these cloning experiments is to isolate many genes from a wide variety of living organisms. The applications of these isolated genes in genetic engineering are varied.

1. Using the clinically important genes for diagnostic purposes. The gene that encodes for one type of hemophilia has been used for this purpose. This field of research is proving to be of invaluable help to people suffering from various diseases like cancer and diabetes and research in this area holds a lot of promise for future generations.

2. Using the isolated genes to isolate similar genes from other organisms. Thus, isolated genes can serve as heterologous probes.

3. Using the isolated gene to derive the nucleic acid sequence. If a partial or complete sequence of the protein that it encodes is available, the gene can be confirmed in this manner. If the protein product is not known then comparing the sequence of the gene with those of the known genes can help derive a function for that gene. Knowing the function of a gene is very important in clinical diagnostic purposes.

4. Isolating a gene that causes disease in a particular crop will help farmers deal with that issue.

Waste water treatment

Wastewater treatment is the process that removes contaminants from sewage. In addition to physical and chemical processing, biological methods are used to clean the wastewater and make it suitable for release back into the environment. Indigenous bacteria can be used to remove biological matter dissolved in the water. Activated sludge is the process in which sewage is aerated to allow the growth of various organisms, collectively known as biological floc, including saprophytic bacteria and protozoan. Advances in biotechnology have led to better management of activated sludge, allowing it to remove the bulk of organic material and the conversion of ammonia to nitrogen gas. These advances include moving bed biological reactors, biological aerated filters, and membrane biological reactors.

Food safety

Food borne illnesses results from the consumption of food contaminated with pathogenic bacteria, viruses, parasites, or other biological toxins. The best preventative measure is proper controls and hygiene during food preparation. However, advances in biotechnology now mean that screening of food for toxins and pathogens can be done more easily. For instance, a fluorescent optical biosensor is currently being developed to measure antibiotic contaminants in milk. The biosensor measures reactions taking place on an antibody-coated probe in contact with the milk products. As testing becomes faster and cheaper, a larger percentage of the food we eat can be screened. Better biological testing methods may also help us identify pathogens that are currently unknown; the cause of approximately 60% of food borne illness is still not known.

-understanding that advancement in technology could caused unforseen issues.

-problem → potential solutions → solutions → evaluate solution → analyze reports

COMPETENCY 3.0   UNDERSTAND THE PROCESS OF SCIENTIFIC INQUIRY AND THE ROLE OF OBSERVATION, EXPERIMENTATION, AND COMMUNICATION IN EXPLAINING NATURAL PHENOMENA

Skill 3.1   Analyzing processes by which new scientific knowledge and hypotheses are generated

The process of scientific inquiry involves questioning, experimentation, and drawing conclusions.

The steps involved in this important process are:

- Observing
- Identifying a problem
- Gathering information/research
- Hypothesizing
- Experimental design, which includes identifying controls, constants, independent and dependent variables
- Conducting experiment and repeating the experiment for validity
- Interpreting analyzing and evaluating data
- Drawing conclusions
- Communicating conclusions

Scientific inquiry starts with observation of a system or phenomenon. After observing, scientists form a question, which starts with "why" or "how". To answer these questions, experimentation is necessary. Between observation and experimentation, there are three more important steps. These are: gathering information (or researching the problem), forming a hypothesis, and designing the experiment.

Designing an experiment is very important since it involves identifying controls, constants, independent variables, and dependent variables. A control or standard is something we compare our results to at the end of the experiment. It is a reference point for comparison. Constants are the factors we remain unchanged in an experiment and do not affect the final results. The independent variable is the factor we change in an experiment (e.g., temperature). It is the focus of our experiment. Finally, dependent variables are the factors that change because of the independent variable (e.g., growth rate at different temperatures). Dependent variables are the factors that we measure in an experiment.

After we complete the experiment, we repeat it and graphically present the results. We then analyze the results and draw conclusions.

It is the responsibility of the scientists to share the knowledge they obtain through their research. After drawing a conclusion, the final step is communication. In this age, much emphasis is put on the means and the method of communication. The conclusions must be communicated by clearly describing the information using accurate data, visual presentations (e.g. bar/line/pie graphs), tables/charts, diagrams, artwork, and other appropriate media such as power point presentations. Scientists should use modern technology whenever it is necessary. The method of communication must be suitable to the audience.

Written communication is as important as oral communication. Written communication skills are essential for submitting research papers to scientific journals, newspapers, and other magazines.

Skill 3.2 Analyzing ethical issues related to the process of scientific research and reporting

Scientists should show good conduct in their scientific pursuits. Conduct here refers to all aspects of scientific activity including experimentation, testing, education, data evaluation, data analysis, data storing, and peer review.

To understand scientific ethics, we need to have a clear understanding of general ethics. Ethics is a system of public, general rules that serve to guide human conduct. The rules are general because they apply to all people at all times and they are public because they are not secret codes or practices.

Philosophers have provided a number of moral theories to justify moral rules ranging from utilitarianism to social contract theory. Utilitarianism, proposed by Mozi, a Chinese philosopher who lived from 471-381 BC, is a theory of ethics based on the determination of what is best for the greatest number of people. Kantianism, a theory proposed by Immanuel Kant, a German philosopher who lived from 1724-1804, ascribes intrinsic value to rational beings. Kantianism is the philosophical foundation of contemporary human rights. Social contract theory, on the other hand, is a view of the ancient Greeks, which states that a person's moral and/or political obligations are dependent upon a contract or societal agreement.

The following are some of the guiding principles of scientific ethics:

1. Scientific Honesty: refrain from fabricating or misinterpreting data for personal gain
2. Caution: avoid errors and sloppiness in all scientific experimentation
3. Credit: give credit where credit is due and do not copy
4. Responsibility: report reliable information to the public and do not mislead in the name of science
5. Freedom: freedom to criticize old ideas, question new research, and conduct independent research.

We could add many more principles to this list. Though these principles seem straightforward and clear, it is very difficult to put them into practice since different people can be interpret them in different ways. Nevertheless, there is no excuse for scientists to overlook these guiding principles of scientific ethics.

Skill 3.3   Evaluating the appropriateness of a specified experimental design to test a hypothesis

Scientists propose and perform experiments with the sole objective of testing a hypothesis. The aforementioned scientist conducts an experiment with the following characteristics. He has two rows each set up with four stations. The first row has a piece of tile as the base at each station. The second row has a piece of linoleum as the base at each station. The scientist has eight eggs and is prepared to drop one over each station. What is he testing? He is trying to answer whether or not the egg is more likely to break when dropped over one material as opposed to the other. His hypothesis might have been: The egg will be less likely to break when dropped on linoleum.

Skill 3.4   Demonstrating an ability to design a hypothesis-testing inquiry experiment

Suppose our junior scientist wants address his initial question, "If you had carried the eggs in their cardboard container, would they have broken if dropped?" A sensible hypothesis to this question would be that an egg would be less likely to break if it was dropped in its cardboard container, than if it were unprotected. Because reproducibility is important, we need to set up multiple identical stations, or use the same station for repeatedly conducting the same experiment. Either way it is critical that everything is identical. If the scientist wants to study the break rate for one egg in the container, then it needs to be just one egg dropped each time in an identical way. The investigator should systematically walk to each station and drop an egg over each station and record the results. The first four times, the egg should be dropped without enclosing it in a cardboard carton. This is the control. It is a recreation of what happened accidentally in the kitchen and one would expect the results to be the same; an egg dropped onto tile will break. The next four times, the egg should be dropped nestled within its original, store-manufactured, cardboard container. One would expect that the egg would not break, or would break less often under these conditions.

When designing an experiment, one needs to clearly define what one is testing. One also needs to consider the question asked. The more limited the question, the easier it is to set up an experiment to answer it.

Theoretically, if an egg were dropped, the egg would be safest when dropped in a protective carton over a soft surface. However, one should not measure multiple variables at once. Studying multiple variables at once makes the results difficult to analyze. How would the investigator discern which variable was responsible for the result? When evaluating experimental design, make sure to look at the number of variables, how clearly they were defined, and how accurately they were measured. Also consider if the experiment was applicable. Did it make sense and address the hypothesis?

COMPETENCY 4.0 **UNDERSTAND THE PROCESSES OF GATHERING, ORGANIZING, REPORTING, AND INTERPRETING SCIENTIFIC DATA, AND APPLY THIS UNDERSTANDING IN THE CONTEXT OF BIOLOGICAL INVESTIGATIONS**

Science is a body of knowledge systematically derived from study, observations, and experimentation. Its goal is to identify and establish principles and theories that may be applied to solve problems. Pseudoscience, on the other hand, is a belief that is not supported by hard evidence. In other words, there is no scientific methodology or application. Some classic examples of pseudoscience include witchcraft, alien encounters, or any topic explained by hearsay.

Scientific experimentation must be repeatable. Experimentation results in theories that can be disproved and changed. Science depends on communication, agreement, and disagreement among scientists. It is composed of theories, laws, and hypotheses.

Theory - A statement of principles or relationships relating to a natural event or phenomenon, which have been verified and accepted.

Law - An explanation of events that occur with uniformity under the same conditions (e.g., laws of nature, law of gravitation).

Hypothesis - An unproved theory or educated guess followed by research to best explain a phenomena. A theory is a proven hypothesis.

Science is limited by the available technology. An example of this would be the relationship between the discovery of the cell and the invention of the microscope. As our technology improves, more hypotheses will become theories and possibly laws. Data collection methods also limit scientific inquiry. Data may be interpreted differently on different occasions. Limitations of scientific methodology produce explanations that change as new technologies emerge.

The first step in scientific inquiry is posing a question. Next, a hypothesis is formed to provide a plausible explanation. An experiment is then proposed and performed to test this hypothesis. A comparison between the predicted and observed results is the next step. Conclusions are then formed and it is determined whether the hypothesis is correct or incorrect. If incorrect, the next step is to form a new hypothesis and repeat the process.

Skill 4.1 **Evaluating the appropriateness of a given method or procedure for collecting data for a specified purpose**

Drawing conclusions is very interesting and important. In order to draw conclusions, we need to study the data at hand. The data tell us whether or not the hypothesis is correct. If the hypothesis is not correct, we must formulate another hypothesis and conduct another experiment.

If we test a hypothesis and the results are consistent in further experimentation, we can formulate a theory. A theory is a hypothesis, which different scientists have tested repeatedly and produced the same results. A theory has more validity because we can use it to predict future events.

Scientific inquiries should end in formulating an explanation or model. Models should be physical, conceptual, and mathematical. The process of drawing conclusions generates a lot of discussion and arguments. There may be several possible explanations for any given sets of results, though not all of them are reasonable. Properly collecting data yields information that appropriately answers the original question. For example, one wouldn't try to use a graduated cylinder to measure mass, nor would one use a ruler to measure a microscopic item.

Utilizing appropriate measuring devices, using proper units, and careful mathematics will provide strong results. Carefully evaluating and analyzing the data creates a reasonable conclusion.

Skill 4.2 Selecting an appropriate and effective graphic representation (e.g., graph, table, diagram) for organizing, reporting, and analyzing given experimental data

The type of graphic representation used to display observations depends on the type of data collected. A **bar graph** or **histogram** compares different items and helps make comparisons based on the data. To use a **bar graph**, determine the scale used for the graph. Then determine the length of each bar on the graph to determine the amount or number of each category of data.

Example: From the bar graph below, determine the percentage of students eligible for athletic programs, which require that students maintain C or better grades.

quantitative

Solution:

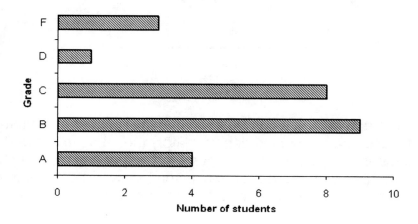

From the bar graph, we can see that 4 students received an A, 9 received a B, 8 received a C, 1 received a D, and 3 students received an F. The total number of students is 4 + 9 + 8 + 1 + 3 = 25 students. Of these, 4 + 9 + 8 = 21 received A, B, or C grades. The percentage of students eligible for athletic programs is 21/25, or 84%.

Line graphs compare different sets of related data and help predict data. For example, a line graph could compare the rate of activity of different enzymes at varying temperatures.

Example: Graph the following information using a line graph and use it to identify trends in National Merit Finalists at Central and Wilson High Schools.

National Merit Finalists

	90-91	91-92	92-93	93-94	94-95	95-96
Central	3	5	1	4	6	8
Wilson	4	2	3	2	3	2

Solution: To make a **line graph**, determine appropriate scales for both the vertical and horizontal axes based on the information to be graphed. Describe what each axis represents and mark the scale periodically on each axis. Graph the individual points of the graph and connect the points on the graph from left to right.

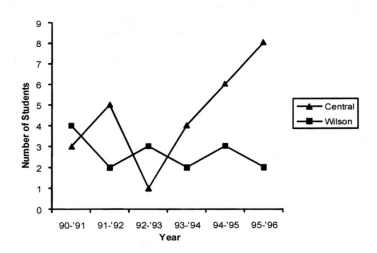

From the line graph above, it can be seen that the number of National Merit Finalists at Wilson has remained approximately the same from 1990 to 1996, while the number of National Merit Finalists at Central has most likely increased in recent years. Because there is some fluctuation over the years at Central, a few more years of data are needed to confirm this trend.

A **pie chart** is useful when organizing data as part of a whole. For example, a pie chart could display the percent of time students spend on various after school activities. To make a **circle** or **pie graph**, total all the information that is to be included on the graph. Determine the central angle to be used for each sector of the graph using the following formula:

percent of total x 360° = degrees in central angle

Lay out the central angles to these sizes, label each section and include its percent of the total.

Example: Graph the following information on a circle graph and identify the percentage of monthly expenses going to basic living expenses (rent, food, and utilities). Monthly expenses: Rent $400, Food $150, Utilities $75, Clothes $75, Church $100, Misc. $200.

Solution:

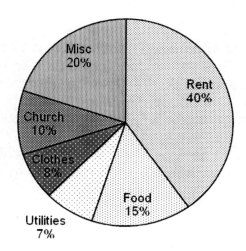

The percentage of expenses going to rent, food, and utilities can be easily added up from the graph: 39% + 15% + 8% = 62%. In addition, a rapid visual assessment of these three sections of the graph (shown in light shading) indicates that the proportion is roughly 2/3.

Scatter plots compare two characteristics of the same group of things or people and usually consist of a large body of data. They show how much one variable is affected by another. The relationship between the two variables is their **correlation**. The closer the data points come to making a straight line when plotted, the closer the correlation.

Example: What conclusions can be drawn from the graph below, correlating salaries of geologists with years of experience?

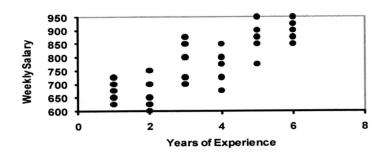

Solution: The weekly salary of geologists is partially correlated with years of experience, since the overall range of salaries paid rises with years of experience. However, other factors are also important, because the ranges overlap substantially among years.

Histograms are used to summarize information from large sets of data that can be naturally grouped into intervals. The vertical axis indicates **frequency** (the number of times any particular data value occurs), and the horizontal axis indicates data values or ranges of data values. The number of data values in any interval is the **frequency of the interval**.

Example: The starting salaries of 13 geologists are shown in the histogram below. What is the most frequent salary offered, and what is the median salary?

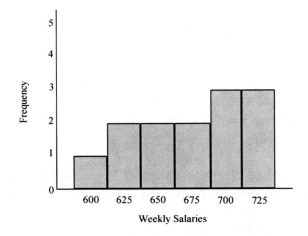

Solution: The most frequent salaries offered are $700/week and $725/week, which occur with equal frequency. To find the median starting salary, translate the above data into a data set for all 13 geologists, ordered from least to most money as follows: {600, 625, 625, 650, 650, 675, 675, 700, 700, 700, 725, 725, and 725}. The median is the middle value from this set, or $675.

Skill 4.3   Demonstrating the ability to appropriately set up and label graphs with dependent and independent variables

The procedure used to obtain data is important to the outcome. Experiments consist of **controls** and **variables**. A control is the experiment run under normal, non-manipulated conditions.

A variable is a factor or condition the scientist manipulates. In biology, the variable may be light, temperature, pH, time, etc. Scientists can use the differences in tested variables to make predictions or form hypotheses. Only one variable should be tested at a time. In other words, one would not alter both the temperature and pH of the experimental subject.

An **independent variable** is one is the researcher directly changes or manipulates. This could be the amount of light given to a plant or the temperature at which bacteria is grown. The **dependent variable** is the factor that changes due to the influence of the independent variable.

Skill 4.4   Applying procedures and criteria for reporting experimental protocols and data (e.g., use of statistical tests)

The knowledge and use of basic mathematical concepts and skills is a necessary aspect of scientific study. Science depends on data and the manipulation of data requires knowledge of mathematics. Scientists often use basic algebra to solve scientific problems and design experiments. For example, the substitution of variables is a common strategy in experiment design. Also, the ability to determine the equation of a curve is valuable in data manipulation, experimentation, and prediction.

Understanding of basic statistics, graphs and charts, and algebra are of particular importance. In addition, scientists must be able to represent data graphically and interpret graphs and tables.

Scientists must be able to understand and apply the statistical concepts of mean, median, mode, and range to sets of scientific data. Modern science requires the use of skills from a number of disciplines. Statistics is one of those subjects, which is absolutely essential for science.

Mean is the mathematical average of all the items. To calculate the mean, all the items must be added up and divided by the number of items. This is also called the arithmetic mean or more commonly the "**average**".

The **median** depends on whether the number of items is odd or even. If the number is odd, then the median is the value of the item in the middle. This is the value that denotes that the number of items having higher or equal value to that is same as the number of items having equal or lesser value than that. If the number of the items is even, the median is the average of the two items in the middle, such that the number of items having values higher or equal to it is same as the number of items having values equal or less than it.

Mode is the value of the item that occurs the most often, if there are not many items. Bimodal is a situation where there are two items with equal frequency. **Range** is the difference between the maximum and minimum values. The range is the difference between two extreme points on the **distribution curve**.

Skill 4.5   Analyzing relationships between factors (e.g., linear, exponential) as indicated by experimental data

When relating factors in experimental data, graphical representations often indicate that two variables vary predictably and in direct relation to each other. Two such relationships common to biological systems are linear and exponential.

The individual data points on the graph of a linear relationship cluster around a line of best fit. In other words, a relationship is linear if we can sketch a straight line that roughly fits the data points. Consider the following examples of linear and non-linear relationships.

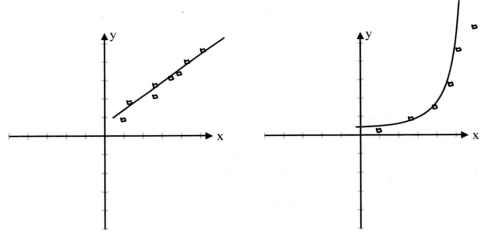

Linear Relationship   Non-Linear Relationship

Note that the non-linear relationship, an exponential relationship in this case, appears linear in parts of the curve. In addition, contrast the preceding graphs to the graph of a data set that shows no relationship between variables.

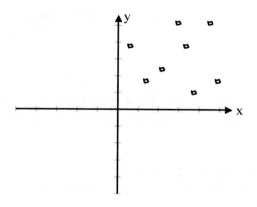

No Relationship

Extrapolation is the process of estimating data points outside a known set of data points. When extrapolating data of a linear relationship, we extend the line of best fit beyond the known values. The extension of the line represents the estimated data points. Extrapolating data is only appropriate if we are relatively certain that the relationship is indeed linear. For example, the death rate of an emerging disease may increase rapidly at first and level off as time goes on. Thus, extrapolating the death rate as if it were linear would yield inappropriately high values at later times.

Similarly, extrapolating certain data in a strictly linear fashion, with no restrictions, may yield obviously inappropriate results. For instance, if the number of plant species in a forest were decreasing with time in a linear fashion, extrapolating the data set to infinity would eventually yield a negative number of species, which is clearly unreasonable.

COMPETENCY 5.0   UNDERSTAND AND APPLY PRINCIPLES AND PROCEDURES OF MEASUREMENT USED IN THE BIOLOGICAL SCIENCES

Skill 5.1   Applying methods of measuring microscopic organisms and structures

The metric system

Science uses the **metric system** (or SI system), as it is accepted worldwide and allows easier comparison among experiments done by scientists around the world.

The meter is the basic metric unit of length. One meter is 1.1 yards. The liter is the basic metric unit of volume. 3.846 liters is 1 gallon. The gram is the basic metric unit of mass. 1000 grams is 2.2 pounds.

The following prefixes define multiples of the basic metric units.

Prefix	Multiplying factor	Prefix	Multiplying factor
deca-	10X the base unit	deci-	1/10 the base unit
hecto-	100X	centi-	1/100
kilo-	1,000X	milli-	1/1,000
mega-	1,000,000X	micro-	1/1,000,000
giga-	1,000,000,000X	nano-	1/1,000,000,000
tera-	1,000,000,000,000X	pico-	1/1,000,000,000,000

Length is the distance from one point to another. The SI unit of length is the meter (m). 1000 meters make 1 kilometer (km).

Volume is the amount of space a substance occupies. The SI unit of volume is the liter (L). 1000 mL makes 1 liter.

Mass is the amount of matter in an object. The SI unit of mass is kilogram (kg). There are 1000 grams in a kilogram.

Time is the period between two events. In SI, time is measured in seconds.

Temperature is a measure of heat in something. The SI scale for measuring temperature is the Kelvin scale (K). In science, temperature is measured in Celsius because it is easier.

Skill 5.2   Evaluating the appropriateness and limitations of units of measurement, measuring devices, or methods of measurement for given situations

The common instrument used for measuring volume is the graduated cylinder. The standard unit of measurement is milliliters (mL). To ensure accurate measurement, it is important to read the liquid in the cylinder at the bottom of the meniscus, the curved surface of the liquid.

The common instrument used in measuring mass is the triple beam balance. The triple beam balance can accurately measure tenths of a gram and can estimate hundredths of a gram.

The ruler and meter stick are the most commonly used instruments for measuring length. As with all scientific measurements, standard units of length are metric. In America, inches and yards are common units of measurement, but they are not appropriate in science. If a ruler has both units, the student should be directed to measure in centimeters, not inches, and meters, not yards. There are 2.54 centimeters in one inch and 1.0936133 yards in one meter.

Skill 5.3   Applying methods of measuring microscopic organisms and structures

Measuring microscopic organisms and structures is basically the same as measuring larger objects. The units, however, are much smaller. The most common units used are:

milli -   1/1,000 of the base unit (e.g., meter)
micro -   1/1,000,000 of the base unit
nano -   1/1,000,000,000 of the base unit
pico -   1/1,000,000,000,000 of the base unit

There are many ways in which errors can creep into measurements. Errors in measurements could occur because of:

- Improper use of instruments used for measuring – weighing, etc.
- Parallax error – improperly positioning the eyes during reading of measurements
- Not using same instruments and methods of measurement during an experiment
- Not using the same source of materials

When scientists use erroneous results to draw conclusions, the conclusions are not reliable. An experiment is valid only when all the constants such as time, place, and method of measurement are strictly controlled.

COMPETENCY 6.0 **UNDERSTAND THE USE OF EQUIPMENT, MATERIALS, CHEMICALS, AND LIVING ORGANISMS IN BIOLOGICAL STUDIES AND THE APPLICATION OF PROCEDURES FOR THEIR PROPER, SAFE, AND LEGAL USE**

Skill 6.1 **Demonstrating knowledge of the appropriate use of given laboratory materials, instruments, and equipment (e.g., indicators, microscopes, centrifuges, spectrophotometers, chromatography equipment)**

Light microscopes are commonly used in high school laboratory experiments. Total magnification is determined by multiplying the magnification of the ocular and the objective lenses. Oculars usually magnify 10X and objective lenses usually magnify 10X on low and 40X on high.

Procedures for the care and use of microscopes include:

- Cleaning all lenses with lens paper only,
- Carrying microscopes with two hands (one on the arm and one on the base),
- Always beginning on low power when focusing before switching to high power,
- Storing microscopes with the low power objective down,
- Always using a cover slip when viewing wet mount slides, *and*
- Bringing the objective down to its lowest position then focusing, moving up to avoid breaking the slide or scratching the lens.

Wet mount slides should be made by placing a drop of water on the specimen and then putting a glass cover slip on top of the drop of water. Dropping the cover slip at a forty-five degree angle will help avoid air bubbles.

Chromatography refers to a set of techniques that are used to separate substances based on their different properties such as size or charge. Paper chromatography uses the principles of capillarity to separate substances such as plant pigments. Molecules of a larger size will move more slowly up the paper, whereas smaller molecules will move more quickly producing lines of pigment.

An **indicator** is any substance used to assist in the classification of another substance. An example of an indicator is litmus paper. Litmus paper is a way to measure whether a substance is acidic or basic. Blue litmus turns pink when an acid is placed on it and pink litmus turns blue when a base is placed on it. pH paper is a more accurate measure of pH, with the paper turning different colors depending on the pH value.

Spectrophotometers measures percent of light at different wavelengths absorbed and transmitted by a pigment solution.

Centrifugation involves spinning substances at a high speed. The more dense part of a solution will settle to the bottom of the test tube, where the lighter material will stay on top. Centrifugation is used to separate blood into blood cells and plasma, with the heavier blood cells settling to the bottom.

Electrophoresis uses electrical charges of molecules to separate them according to their size. The molecules, such as DNA or proteins are pulled through a gel towards either the positive end of the gel box (if the material has a negative charge) or the negative end of the gel box (if the material has a positive charge). DNA is negatively charged and moves towards the positive charge.

One of the most widely used genetic engineering techniques is the **polymerase chain reaction (PCR)**. PCR is a technique in which a piece of DNA can be amplified into billions of copies within a few hours. This process requires a primer to specify the segment to be copied, and an enzyme (usually taq polymerase) to amplify the DNA. PCR has allowed scientists to perform multiple procedures on small amounts of DNA.

Skill 6.2 Applying proper methods for storing, identifying, dispensing, and disposing of chemicals and biological materials

All laboratory solutions should be prepared as directed in the lab manual. Care should be taken to avoid contamination. All glassware should be rinsed thoroughly with distilled water before using and cleaned well after use. All solutions should be made with distilled water as tap water contains dissolved particles that may affect the results of an experiment. Unused solutions should be disposed of according to local disposal procedures.

The "Right to Know Law" covers science teachers who work with potentially hazardous chemicals. Briefly, the law states that employees must be informed of potentially toxic chemicals. An inventory must be made available if requested. The inventory must contain information about the hazards and properties of the chemicals. This inventory is to be checked against the "Substance List". Training must be provided on safe handling and interpretation of the Material Safety Data Sheet.

The following chemicals are potential carcinogens and not allowed in school facilities: Acrylonitrile, Arsenic compounds, Asbestos, Benzidine, Benzene, Cadmium compounds, Chloroform, Chromium compounds, Ethylene oxide, Ortho-toluidine, Nickel powder, and Mercury.

Chemicals should not be stored on bench tops or heat sources. They should be stored in groups based on their reactivity with one another and in protective storage cabinets. All containers within the lab must be labeled. Suspected and known carcinogens must be labeled as such and stored in trays to contain leaks and spills.

Chemical waste should be disposed of in properly labeled containers. Waste should be separated based on its reactivity with other chemicals.

Biological material should never be stored near food or water used for human consumption. All biological material should be appropriately labeled. All blood and body fluids should be put in a well-contained container with a secure lid to prevent leaking. All biological waste should be disposed of in biological hazardous waste bags.

Skill 6.3 **Identifying sources of and interpreting information (e.g., material safety data sheets) regarding the proper, safe, and legal use of equipment, materials, and chemicals**

Material safety data sheets (MSDS) are available for every chemical and biological substance. These are available directly from the distribution company and on the internet at www.msds.com. Before using lab equipment, all lab workers should read and understand the equipment manuals.

MSDS's include information such as physical data (melting point, boiling point, etc.), toxicity, health effects, first aid, reactivity, storage, disposal, protective gear, and spill/leak procedures. These are particularly important if a spill or other accident occurs. You should review a few, commonly available online, and understand the listing procedures. The manuals for equipment used in the lab should be read and understood before use of equipment.

Skill 6.4 **Interpreting guidelines and regulations for the proper and humane procurement and treatment of living organisms in biological studies**

No dissections may be performed on living mammalian vertebrates or birds. Lower order life and invertebrates may be used. Biological experiments may be done with all animals except mammalian vertebrates or birds. No physiological harm may result to the animal. All animals housed and cared for in the school must be handled in a safe and humane manner. Animals are not to remain on school premises during extended vacations unless adequate care is provided. Any instructor who intentionally refuses to comply with the laws may be suspended or dismissed.

Pathogenic organisms must never be used for experimentation. Students should adhere to the following rules at all times when working with microorganisms to avoid accidental contamination:

1. Treat all microorganisms as if they were pathogenic.
2. Maintain sterile conditions at all times.

Dissection and alternatives to dissection

Animals which were not obtained from recognized sources should not be used. Decaying animals or those of unknown origin may harbor pathogens and/or parasites. Specimens should be rinsed before handling. Latex gloves are recommended. If not available, students with sores or scratches should be excused from the activity. Formaldehyde is likely carcinogenic and should be avoided or disposed of according to district regulations. Students objecting to dissections for moral reasons should be given an alternative assignment. For those students who object to performing dissections, interactive dissections are available online at http://frog.edschool.virginia.edu or from software companies. There should be no penalty for those students who refuse to physically perform a dissection.

Skill 6.5   **Applying proper procedures for promoting laboratory safety (e.g., the use of safety goggles, universal health precautions) and responding to accidents and injuries in the biology laboratory**

All science labs should contain the following **safety equipment**.

- Fire blanket that is visible and accessible
- Ground Fault Circuit Interrupters (GFCI) within two feet of water supplies
- Signs designating room exits
- Emergency shower providing a continuous flow of water
- Emergency eye wash station that can be activated by the foot or forearm
- Eye protection for every student
- A means of sanitizing equipment
- Emergency exhaust fans providing ventilation to the outside of the building
- Master cut-off switches for gas, electric, and compressed air; switches must have permanently attached handles. Cut-off switches must be clearly labeled
- An ABC fire extinguisher
- Storage cabinets for flammable materials
- Chemical spill control kit
- Fume hood with a motor that is spark proof
- Protective laboratory aprons made of flame retardant material
- Signs that will alert of potential hazardous conditions

- Labeled containers for broken glassware, flammables, corrosives, and waste

Students should wear safety goggles when performing dissections, heating, or while using acids and bases. Hair should always be tied back and objects should never be placed in the mouth. Food should not be consumed while in the laboratory. Hands should always be washed before and after laboratory experiments. In case of an accident, eye washes and showers should be used for eye contamination or a chemical spill that covers the student's body. Small chemical spills should only be contained and cleaned by the teacher.

Kitty litter or a chemical spill kit should be used to clean a spill. For large spills, the school administration and the local fire department should be notified. Biological spills should only be handled by the teacher. Contamination with biological waste can be cleaned by using bleach when appropriate. Accidents and injuries should always be reported to the school administration and local health facilities. The severity of the accident or injury will determine the course of action.

It is the responsibility of the teacher to provide a safe environment for his or her students. Proper supervision greatly reduces the risk of injury and a teacher should never leave a class for any reason without providing alternate supervision. After an accident, two factors are considered, **foreseeability** and **negligence**. Foreseeability is the anticipation that an event may occur under certain circumstances. Negligence is the failure to exercise ordinary or reasonable care. Safety procedures should be a part of the science curriculum and a well managed classroom is important to avoid potential lawsuits.

SUBAREA II.   **CELL BIOLOGY AND BIOCHEMISTRY**

COMPETENCY 7.0   UNDERSTAND CELL STRUCTURE AND FUNCTION, THE DYNAMIC NATURE OF CELLS, AND THE UNIQUENESS OF DIFFERENT TYPES OF CELLS

Skill 7.1   Comparing and contrasting the cellular structures and functions of archaea, prokaryotes, and eukaryotes

The cell is the basic unit of all living things. There are three types of cells: prokaryotic, eukaryotic, and archaea. Archaea have some similarities with prokaryotes, but are as distantly related to prokaryotes as prokaryotes are to eukaryotes.

PROKARYOTES

~Single celled organism

Prokaryotes consist only of bacteria and cyanobacteria (formerly known as blue-green algae). The diagram below shows the classification of prokaryotes.

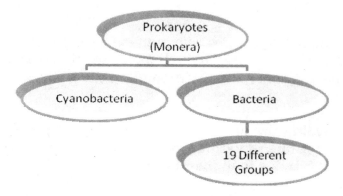

Bacterial cells have no defined nucleus or nuclear membrane. The DNA, RNA, and ribosomes float freely within the cell. The cytoplasm has a single chromosome condensed to form a **nucleoid**. Prokaryotes have a thick cell wall made up of amino sugars (glycoproteins) that provides protection, gives the cell shape, and keeps the cell from bursting. The antibiotic penicillin targets the **cell wall** of bacteria. Penicillin works by disrupting the cell wall, thus killing the cell.

The cell wall surrounds the **cell membrane** (plasma membrane). The cell membrane consists of a lipid bilayer that controls the passage of molecules in and out of the cell. Some prokaryotes have a capsule made of polysaccharides that surrounds the cell wall for extra protection from higher organisms.

Many bacterial cells have appendages used for movement called **flagella**. Some cells also have **pili**, which are a protein strand used for attachment. Pili may also be used for sexual conjugation (where bacterial cells exchange DNA).

Prokaryotes are the most numerous and widespread organisms on earth. Bacteria were most likely the first cells and date back in the fossil record to 3.5 billion years ago. Their ability to adapt to the environment allows them to thrive in a wide variety of habitats.

EUKARYOTES → *multi cellular organism*

Eukaryotic cells are found in protists, fungi, plants, and animals. Most eukaryotic cells are larger than prokaryotic cells. They contain many organelles, which are membrane bound areas for specific functions. Their cytoplasm contains a cytoskeleton which provides a protein framework for the cell. The cytoplasm also supports the organelles and contains the ions and molecules necessary for cell function. The cytoplasm is contained by the plasma membrane. The plasma membrane allows molecules to pass in and out of the cell. The membrane can bud inward to engulf outside material in a process called endocytosis. Exocytosis is a secretory mechanism, the reverse of endocytosis.

ARCHAEA

There are three kinds of organisms with archaea cells: **methanogens**, obligate anaerobes that produce methane, **halobacteria**, which can live only in concentrated brine solutions, and **thermoacidophiles**, which can live only in acidic hot springs.

Skill 7.2   **Analyzing the primary functions, processes, products, and interactions of various cellular structures (e.g., lysosomes, microtubules, cell membrane)**

The most significant difference between prokaryotes and eukaryotes is that eukaryotes have a **nucleus**. The nucleus is the "brain" of the cell that contains all of the cell's genetic information. The chromosomes consist of chromatin, which are complexes of DNA and proteins. The chromosomes are tightly coiled to conserve space while providing a large surface area. The nucleus is the site of transcription of the DNA into RNA. The **nucleolus** is where ribosomes are made. There is at least one of these dark-staining bodies inside the nucleus of most eukaryotes. The nuclear envelope consists of two membranes separated by a narrow space. The envelope contains many pores that let RNA out of the nucleus.

Ribosomes are the site for protein synthesis. Ribosomes may be free floating in the cytoplasm or attached to the endoplasmic reticulum. There may be up to a half a million ribosomes in a cell, depending on how much protein the cell makes.

The **endoplasmic reticulum (ER)** is folded and has a large surface area. It is the "roadway" of the cell and allows for transport of materials through and out of the cell. There are two types of ER: smooth and rough. Smooth endoplasmic reticula contain no ribosomes on their surface and are the site of lipid synthesis. Rough endoplasmic reticula have ribosomes on their surface and aid in the synthesis of proteins that are membrane bound or destined for secretion.

Many of the products made in the ER proceed to the Golgi apparatus. The **Golgi apparatus** functions to sort, modify, and package molecules that are made in the other parts of the cell (like the ER). These molecules are either sent out of the cell or to other organelles within the cell.

Lysosomes are found mainly in animal cells. These contain digestive enzymes that break down food, unnecessary substances, viruses, damaged cell components, and, eventually, the cell itself. It is believed that lysosomes play a role in the aging process.

Mitochondria are large organelles that are the site of cellular respiration, the production of ATP that supplies energy to the cell. Muscle cells have many mitochondria because they use a great deal of energy. Mitochondria have their own DNA, RNA, and ribosomes and are capable of reproducing by binary fission if there is a great demand for additional energy. Mitochondria have two membranes: a smooth outer membrane and a folded inner membrane. The folds inside the mitochondria are called cristae. They provide a large surface area for cellular respiration to occur.

Plastids are found only in photosynthetic organisms. They are similar to the mitochondria due to the double membrane structure. They also have their own DNA, RNA, and ribosomes and can reproduce if the need for the increased capture of sunlight becomes necessary. There are several types of plastids. **Chloroplasts** are the sight of photosynthesis. The stroma is the chloroplast's inner membrane space. The stoma encloses sacs called thylakoids that contain the photosynthetic pigment chlorophyll. The chlorophyll traps sunlight inside the thylakoid to generate ATP which is used in the stroma to produce carbohydrates and other products. The **chromoplasts** make and store yellow and orange pigments. They provide color to leaves, flowers, and fruits. The **amyloplasts** store starch and are used as a food reserve. They are abundant in roots like potatoes.

The Endosymbiotic Theory states that mitochondria and chloroplasts were once free living and possibly evolved from prokaryotic cells. At some point in our evolutionary history, they entered the eukaryotic cell and maintained a symbiotic relationship with the cell, with both the cell and organelle benefiting from the relationship. The fact that they both have their own DNA, RNA, ribosomes, and are capable of reproduction supports this theory.

Found only in plant cells, the **cell wall** is composed of cellulose and fibers. It is thick enough for support and protection, yet porous enough to allow water and dissolved substances to enter. **Vacuoles** are found mostly in plant cells. They hold stored food and pigments. Their large size allows them to fill with water in order to provide turgor pressure. Lack of turgor pressure causes a plant to wilt.

The **cytoskeleton**, found in both animal and plant cells, is composed of protein filaments attached to the plasma membrane and organelles. The cytoskeleton provides a framework for the cell and aids in cell movement. Three types of fibers make up the cytoskeleton:

1. **Microtubules** – The largest of the three fibers, they make up cilia and flagella for locomotion. Some examples are sperm cells, cilia that line the fallopian tubes, and tracheal cilia. Centrioles are also composed of microtubules. They aid in cell division to form the spindle fibers that pull the cell apart into two new cells. Centrioles are not found in the cells of higher plants.

2. **Intermediate filaments** – Intermediate in size, they are smaller than microtubules, but larger than microfilaments. They help the cell keep its shape.

3. **Microfilaments** – Smallest of the three fibers, they are made of actin and small amounts of myosin (like in muscle tissue). They function in cell movement like cytoplasmic streaming, endocytosis, and amoeboid movement. This structure pinches the two cells apart after cell division, forming two new cells.

The following is a diagram of a generalized animal cell.

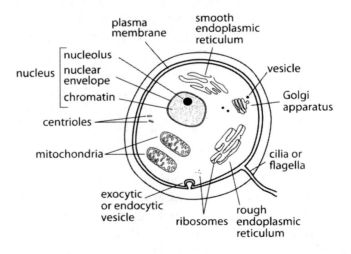

Skill 7.3   Analyzing the importance of active and passive transport processes in maintaining homeostasis in cells and the relationships between these processes and the cellular membranes

In order to understand cellular transport, it is important to know about the structure of the cell membrane. All organisms contain cell membranes because they regulate the flow of materials into and out of the cell. The current model for the cell membrane is the Fluid Mosaic Model, which takes into account the ability of lipids and proteins to move and change places, giving the membrane fluidity.

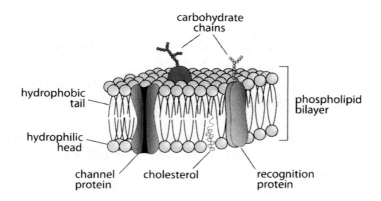

Cell membranes have the following characteristics:

1. They are made of phospholipids which have polar, charged heads with a phosphate group which is hydrophilic (water loving) and two nonpolar lipid tails which are hydrophobic (water fearing). This allows the membrane to orient itself with the polar heads facing the fluid inside and outside the cell and the hydrophobic lipid tails sandwiched in between. Each individual phospholipid is called a micelle.

2. They contain proteins embedded inside (integral proteins) and proteins on the surface (peripheral proteins). These proteins may act as channels for transport, may contain enzymes, may act as receptor sites, may act to stick cells together, or may attach to the cytoskeleton to give the cell shape.

3. They contain cholesterol, which alters the fluidity of the membrane.

4. They contain oligosaccharides (small carbohydrate polymers) on the outside of the membrane. These act as markers that help distinguish one cell from another.

5. They contain receptors made of glycoproteins that can attach to certain molecules, like hormones.

Cell transport is necessary to maintain homeostasis, or balance between the cell and its external environment. Cell membranes are selectively permeable, which is the key to transport. Not all molecules may pass through easily. Some molecules require energy or carrier molecules and may only cross when needed.

Passive transport does not require energy and moves the material with the concentration gradient (high to low). Small molecules may pass through the membrane in this manner. Two examples of passive transport include diffusion and osmosis. **Diffusion** is the ability of molecules to move from areas of high concentration to areas of low concentration. It normally involves small uncharged particles like oxygen. **Osmosis** is simply the diffusion of water across a semi-permeable membrane. Osmosis may cause cells to swell or shrink, depending on the internal and external environments. The following terms are used to describe the relationship of the cell to the environment.

Isotonic - Water concentration is equal inside and outside the cell. Net movement in either direction is basically equal.

Hypertonic - "Hyper" refers to the amount of dissolved particles. The more particles are in a solution, the lower its water concentration. Therefore, when a cell is hypertonic to its environment, there is more water outside the cell than inside. Water will move into the cell and the cell will swell. If the environment is hypertonic to the cell, there is more water inside the cell. Water will move out of the cell and the cell will shrink.

Hypotonic - "Hypo" again refers to the amount of dissolved particles. The fewer particles are in solution, the higher its water concentration. When a cell is hypotonic to its environment, there is more water inside the cell than outside. Water will move out of the cell and the cell will shrink. If the environment is hypotonic to the cell, there is more water outside the cell than inside. Water will move into the cell and the cell will swell.

The **facilitated diffusion** mechanism does not require energy, but does require a carrier protein. An example is insulin, which is needed to carry glucose into the cell.

Active transport requires energy. The energy for this process comes from either ATP or an electrical charge difference. Active transport may move materials either with or against a concentration gradient. Some examples of active transport are:

- Sodium-Potassium pump - maintains an electrical difference across the cell. This is useful in restoring ion balance so nerves can continue to function. It exchanges sodium ions for potassium ions across the plasma membrane in animal cells.

- Stomach acid pump - exports hydrogen ions to lower the pH of the stomach and increase acidity.

- Calcium pumps - actively pump calcium outside of the cell and are important in nerve and muscle transmission.

Active transport involves a membrane potential, which is a charge on the membrane. The charge works like a magnet and may cause transport proteins to alter their shape, thus allowing substances in or out of the cell.

The transport of large molecules depends on the fluidity of the membrane, which is controlled by cholesterol in the membrane. **Exocytosis** is the release of large particles by vesicles fusing with the plasma membrane. In the process of **endocytosis**, the cell takes in macromolecules and particulate matter by forming vesicles derived from the plasma membrane. There are three types of endocytosis in animal cells. **Phagocytosis** is when a particle is engulfed by pseudopodia and packaged in a vacuole. In **pinocytosis**, the cell takes in extracellular fluid in small vesicles. **Receptor-mediated endocytosis** is when the membrane vesicles bud inward to allow a cell to take in large amounts of certain substances. The vesicles have proteins with receptors that are specific for the substance.

-Passive transport: no energy; moves material from high to low
-Active transport: needs energy; from ATP or the charges

COMPETENCY 8.0   **UNDERSTAND CHEMISTRY AND BIOCHEMISTRY TO ANALYZE THE ROLE OF BIOLOGICALLY IMPORTANT ELEMENTS AND COMPOUNDS IN LIVING ORGANISMS**

Skill 8.1   **Comparing and contrasting hydrogen, ionic, and covalent bonds and the conditions under which these bonds form and break apart**

Chemical bonds form when atoms with incomplete valence shells share or completely transfer their valence electrons. There are three types of chemical bonds; covalent bonds, ionic bonds, and hydrogen bonds.

Covalent bonding is the sharing of a pair of valence electrons by two atoms. A simple example of this is two hydrogen atoms. Each hydrogen atom has one valence electron in its outer shell, therefore the two hydrogen atoms come together to share their electrons. Some atoms share two pairs of valence electrons, like two oxygen atoms. This is a double covalent bond.

Electronegativity describes the ability of an atom to attract electrons toward itself. The greater the electronegativity of an atom, the more it pulls the shared electrons towards itself. Electronegativity of the atoms determines whether the bond is polar or nonpolar. In **nonpolar covalent bonds**, the electrons are shared equally, thus the electronegativity of the two atoms is the same. This type of bonding usually occurs between two of the same atoms. A **polar covalent bond** forms when different atoms join, as in hydrogen and oxygen to create water. In this case, oxygen is more electronegative than hydrogen so the oxygen pulls the hydrogen electrons toward itself.

Ionic bonds form when one electron is stripped away from its atom to join another atom. An example of this is sodium chloride (NaCl). A single electron on the outer shell of sodium joins the chloride atom with seven electrons in its outer shell. The sodium now has a +1 charge and the chloride now has a -1 charge. The charges attract each other to form an ionic bond. Ionic compounds are called salts. In a dry salt crystal, the bond is so strong it requires a great deal of strength to break it apart. However, if the salt crystal is placed in water, the bond will dissolve easily as the attraction between the two atoms decreases.

The weakest of the three bonds is the **hydrogen bond**. A hydrogen bond forms when one electronegative atom shares a hydrogen atom with another electronegative atom. An example of a hydrogen bond is a water molecule (H_2O) bonding with an ammonia molecule (NH_3). The H^+ atom of the water molecule attracts the negatively charged nitrogen in a weak bond. Weak hydrogen bonds are beneficial because they can briefly form, the atoms can respond to one another, and then break apart allowing formation of new bonds. Hydrogen bonding plays a very important role in the chemistry of life.

Skill 8.2 **Relating the structure and function of carbohydrates, lipids, proteins (e.g., level of structure), nucleic acids, and inorganic compounds to cellular activities**

A compound consists of two or more elements. There are four major chemical compounds found in the cells and bodies of living things. These are carbohydrates, lipids, proteins, and nucleic acids.

Monomers are the simplest unit of structure. **Monomers** combine to form **polymers**, or long chains, making a large variety of molecules. Monomers combine through the process of condensation reactions (also called dehydration synthesis). In this process, one molecule of water is removed between each of the adjoining molecules. In order to break the molecules apart in a polymer, water molecules are added between monomers, thus breaking the bonds between them. This process is called hydrolysis.

Carbohydrates contain a ratio of two hydrogen atoms for each carbon and oxygen $(CH_2O)_n$. Carbohydrates include sugars and starches. They function in the release of energy. **Monosaccharide's** are the simplest sugars and include glucose, fructose, and galactose. They are the major nutrients for cells. In cellular respiration, the cells extract the energy from glucose molecules. **Disaccharides** are made by joining two monosaccharides by condensation to form a glycosidic linkage (covalent bond between two monosaccharides). Maltose is the combination of two glucose molecules, lactose is the combination of glucose and galactose, and sucrose is the combination of glucose and fructose. **Polysaccharides** consist of many monomers joined together and may be structural or provide energy storage for the cell. As energy stores, polysaccharides are hydrolyzed as needed to provide sugar for cells. Examples of polysaccharides include starch, glycogen, cellulose, and chitin.

> **Starch** - major energy storage molecule in plants. It is a polymer consisting of glucose monomers.
>
> **Glycogen** - major energy storage molecule in animals. It is made up of many glucose molecules.
>
> **Cellulose** - found in plant cell walls, its function is structural. Many animals lack the enzymes necessary to hydrolyze cellulose, so it simply adds bulk (fiber) to the diet.
>
> **Chitin** - found in the exoskeleton of arthropods and fungi. Chitin contains an amino sugar (glycoprotein).

Lipids are composed of glycerol (an alcohol) and three fatty acids. Lipids are **hydrophobic** (water fearing) and will not mix with water. There are three important families of lipids; fats, phospholipids and steroids.

Fats consist of glycerol (alcohol) and three fatty acids. Fatty acids are long carbon skeletons. The nonpolar carbon-hydrogen bonds in the tails of fatty acids are highly hydrophobic. Fats are solids at room temperature and come from animal sources (e.g., butter and lard).

Phospholipids are a vital component in cell membranes. In a phospholipid, one or two fatty acids are replaced by a phosphate group linked to a nitrogen group. They consist of a **polar** (charged) head that is hydrophilic (water loving) and a **nonpolar** (uncharged) tail which is hydrophobic. This allows the membrane to orient itself with the polar heads facing the interstitial fluid found outside the cell and the nonpolar tails facing the internal fluid of the cell.

Steroids are insoluble and are composed of a carbon skeleton consisting of four inter-connected rings. An important steroid is cholesterol, which is the precursor from which other steroids are synthesized. Hormones, including cortisone, testosterone, estrogen, and progesterone, are steroids. Their insolubility keeps them from dissolving in body fluids.

Proteins comprise about fifty percent of the dry weight of animals and bacteria. Proteins function in structure and support (e.g., connective tissue, hair, feathers, and quills), storage of amino acids (e.g., albumin in eggs and casein in milk), transport of substances (e.g. hemoglobin), coordination of body activities (e.g. insulin), signal transduction (e.g. membrane receptor proteins), contraction (e.g., muscles, cilia, and flagella), body defense (e.g. antibodies), and as enzymes to speed up chemical reactions.

All proteins are made of twenty **amino acids**. An amino acid contains an amino group and an acid group. The radical group varies and defines the amino acid. Amino acids form through condensation reactions that remove water. The bond formed between two amino acids is called a peptide bond. Polymers of amino acids are called polypeptide chains. An analogy can be drawn between the twenty amino acids and the alphabet. We can form millions of words using an alphabet of only twenty-six letters. Similarly, organisms can create many different proteins using the twenty amino acids. This results in the formation of many different proteins, whose structure typically defines its function.

There are four levels of protein structure: primary, secondary, tertiary, and quaternary. **Primary structure** is the protein's unique sequence of amino acids. A slight change in primary structure can affect a protein's conformation and its ability to function. **Secondary structure** is the coils and folds of polypeptide chains. The coils and folds are the result of hydrogen bonds along the polypeptide backbone. The secondary structure is either in the form of an alpha helix or a pleated beta sheet. The alpha helix is a coil held together by hydrogen bonds. A beta pleated sheet is the polypeptide chain folding back and forth. The hydrogen bonds between parallel regions hold it together. **Tertiary structure** results from bonds between the side chains of the amino acids. For example, disulfide bridges form when two sulfhydryl groups on the amino acids form a strong covalent bond. **Quaternary structure** is the overall structure of the protein from the aggregation of two or more polypeptide chains. An example of this is hemoglobin. Hemoglobin consists of two kinds of polypeptide chains.

Nucleic acids consist of DNA (deoxyribonucleic acid) and RNA (ribonucleic acid).

Nucleic acids contain the code for the amino acid sequence of proteins and the instructions for replicating. The monomer of nucleic acids is a nucleotide. A nucleotide consists of a 5-carbon sugar (deoxyribose in DNA, ribose in RNA), a phosphate group, and a nitrogenous base. The base sequence is the code or the instructions. There are five bases: adenine, thymine, cytosine, guanine, and uracil. Uracil is found only in RNA and replaces thymine. The following provides a summary of nucleic acid structure:

	SUGAR	PHOSPHATE	BASES
DNA	Deoxyribose	Present	adenine, **thymine**, cytosine, guanine
RNA	Ribose	Present	adenine, **uracil**, cytosine, guanine

Due to the molecular structure, adenine will always pair with thymine in DNA or uracil in RNA. Cytosine always pairs with guanine in both DNA and RNA.

This allows for the symmetry of the DNA molecule seen below.

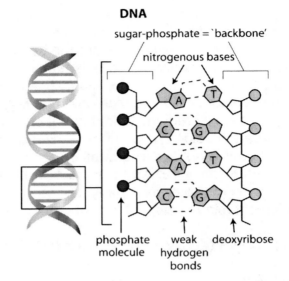

DNA

Adenine and thymine (or uracil) are linked by two covalent bonds and cytosine and guanine are linked by three covalent bonds. Guanine and cytosine are harder to break apart than thymine (uracil) and adenine because of the greater number of bonds between the bases. The double-stranded DNA molecule forms a double helix, or twisted ladder, shape.

Skill 8.3   Analyzing the properties of water and the significance of these properties to living organisms

Water is necessary for life. The unique properties of water are due to its molecular structure. Water is an important solvent in biological compounds. Water is a polar substance. This means it is formed by covalent bonds that make it electrically lopsided. Water molecules are attracted to other water molecules due to this electrical attraction and this allows for two important properties: adhesion and cohesion.

Adhesion is when water sticks to other substances like the xylem of a stem, which aids the water in traveling up the stem to the leaves.

Cohesion is the ability of water molecules to stick to each other by hydrogen bonding. This allows for surface tension on a body of water, or capillarity, which allows water to move through vessels. Surface tension is how difficult it is to stretch or break the surface of a liquid. Cohesion allows water to move against gravity.

There are several other important properties of water. Water is a good solvent. An aqueous solution is one in which water is the solvent. It provides a medium for chemical reactions to occur. Water has a high specific heat of 1 calorie per gram per degree Celsius, allowing it to cool and warm slowly, allowing organisms to adapt to temperature changes. Water has a high boiling point; thus, it is a good coolant. Its ability to evaporate stabilizes the environment and allows organisms to maintain body temperature. Water has a high freezing point and a lower density as a solid than as a liquid. Water is most dense at four degrees Celsius. This allows ice to float on top of water so a whole body of water does not freeze during the winter. Because of this property of water, aquatic animals can survive the winter.

Skill 8.4 Demonstrating an understanding of pH chemistry in biological systems

The pH scale tells how acidic or basic a solution is. An acid is a solution that increases the hydrogen ion concentration of a solution. An example is hydrochloric acid (HCl). A base is a solution that reduces the hydrogen ion concentration. An example is sodium hydroxide (NaOH).

The pH scale ranges from zero to fourteen. Seven is a neutral solution. This is the pH of pure water. A pH between zero and 6.9 is acidic. Stomach acid has a pH of 2.0. A pH between 7.1 and 14 is basic. Common household bleach has a pH of 12.

The internal pH of most living organisms is close to 7. Human blood has a pH of 7.4. Variation from this neutral pH can be harmful to the living organism. Biological fluids resist pH variation due to buffers that minimize the effects of H^+ and OH^- concentrations. A buffer accepts or donates H+ ions from or to the solution when they are in excess or depleted.

The pH of a substance has a dramatic effect on the environment as well. Acids greatly affect the environment. Acidic precipitation is rain, snow, or fog with a pH less than 5.6. Acidic precipitation is caused by sulfur oxides and nitrogen oxides in the environment that react with water in the air to form acids that fall down to earth as precipitation. A change in pH in the environment can affect the solubility of minerals in the soil, which causes a delay in forest growth.

Skill 8.5 Analyze the structure and function of enzymes and factors that affect the rate of reactions

Enzymes act as biological catalysts that speed up reactions. Enzymes are the most diverse type of protein. They are not used up in a reaction and are recyclable. Each enzyme is specific for a single reaction. Enzymes act on a substrate. The substrate is the material to be broken down or put back together.

Most enzymes end in the suffix -ase (lipase, amylase). The prefix is the substrate being acted on (lipids, sugars).

Enzyme

Substrate $\longrightarrow$ Product

The active site is the region of the enzyme that binds to the substrate. There are two theories for how the active site functions. The **lock and key theory** states that the shape of the enzyme is specific because it fits into the substrate like a key fits into a lock. In this theory, the enzyme holds molecules close together so reactions can easily occur. The **Induced fit theory** states that an enzyme can stretch and bend to fit the substrate. This is the most accepted theory.

Many factors can affect enzyme activity. Temperature and pH are two such factors. The temperature can affect the rate of reaction of an enzyme. The optimal pH for enzymes is between 6 and 8, with a few enzymes whose optimal pH falls outside of this range.

Cofactors aid in the enzyme's function. Cofactors may be inorganic or organic. Organic cofactors are known as coenzymes. Vitamins are examples of coenzymes. Some chemicals can inhibit an enzyme's function. **Competitive inhibitors** block the substrate from entering the active site of the enzyme to reduce productivity. **Noncompetitive inhibitors** bind to a location on the enzyme that is different from the active site and interrupts substrate binding. In most cases, noncompetitive inhibitors alter the shape of the enzyme. An **allosteric enzyme** can exist in two shapes; they are active in one form and inactive in the other. Overactive enzymes may cause metabolic diseases.

COMPETENCY 9.0   **UNDERSTAND THE RAW MATERIALS, PRODUCTS, AND SIGNIFICANCE OF PHOTOSYNTHESIS AND CELLULAR RESPIRATION AND THE RELATIONSHIPS OF THESE PROCESSES TO CELL STRUCTURE AND FUNCTION**

Skill 9.1   **Recognizing the significance of photosynthesis and respiration to living organisms**

Cellular respiration is the metabolic pathway in which food (e.g. glucose) is broken down to produce energy in the form of ATP. Both plants and animals utilize respiration to create energy for metabolism. In respiration, energy is released by the transfer of electrons in a process known as an **oxidation-reduction (redox)** reaction. The oxidation phase of this reaction is the loss of an electron and the reduction phase is the gain of an electron. Redox reactions are important for all stages of respiration. Specific pathways are discussed in the following sections.

Skill 9.2   **Identifying the overall chemical equations for the processes of respiration and photosynthesis**

Glycolysis is the first step in respiration. It occurs in the cytoplasm of the cell and does not require oxygen. Each of the ten stages of glycolysis are catalyzed by a specific enzyme. The following is a summary of these stages.

In the first stage the reactant is glucose. For energy to be released from glucose, it must be converted to a reactive compound. This conversion occurs through the phosphorylation of a molecule of glucose by the use of two molecules of ATP. This is an investment of energy by the cell. The 6-carbon product, called fructose -1,6- bisphosphate, breaks into two 3-carbon molecules of sugar. A phosphate group is added to each sugar molecule and hydrogen atoms are removed. Hydrogen is picked up by NAD^+ (a vitamin). Since there are two sugar molecules, two molecules of NADH are formed. The reduction (addition of hydrogen) of NAD allows the potential for energy transfer.

As the phosphate bonds are broken, ATP is produced. Two ATP molecules are generated as each original 3-carbon sugar molecule is converted to pyruvic acid (pyruvate). A total of four ATP molecules are made in the four stages. Since two molecules of ATP were needed to start the reaction in stage 1, there is a net gain of two ATP molecules at the end of glycolysis. This accounts for only two percent of the total energy in a molecule of glucose.

Beginning with pyruvate, which was the end product of glycolysis, the following steps occur before entering the **Krebs cycle**.

1. Pyruvic acid is changed to acetyl-CoA (coenzyme A). This is a 3-carbon pyruvic acid molecule, which has lost one molecule of carbon dioxide (CO_2) to become a 2-carbon acetyl group. Pyruvic acid loses a hydrogen atom to NAD^+, which is reduced to NADH.

2. Acetyl CoA enters the Krebs cycle. For each molecule of glucose it started with, two molecules of Acetyl CoA enter the Krebs cycle (one for each molecule of pyruvic acid formed in glycolysis).

The **Krebs cycle** (also known as the citric acid cycle), occurs in four major steps. First, the 2-carbon acetyl CoA combines with a 4-carbon molecule to form a 6-carbon molecule of citric acid. Next, two carbons are lost as carbon dioxide (CO_2) and a 4-carbon molecule is formed to become available to join with CoA to form citric acid again. Since we started with two molecules of CoA, two turns of the Krebs cycle are necessary to process the original molecule of glucose. In the third step, eight hydrogen atoms are released and picked up by FAD and NAD (vitamins and electron carriers). Lastly, for each molecule of CoA (remember there were two to start with) you get:

> 3 molecules of NADH x 2 cycles
> 1 molecule of $FADH_2$ x 2 cycles
> 1 molecule of ATP x 2 cycles

Therefore, this completes the breakdown of glucose. At this point, a total of four molecules of ATP have been made, two from glycolysis and one from each of the two turns of the Krebs cycle. Six molecules of carbon dioxide have been released, two prior to entering the Krebs cycle and two for each of the two turns of the Krebs cycle. Twelve carrier molecules have been made, ten NADH and two $FADH_2$. These carrier molecules will carry electrons to the electron transport chain.

ATP is made by substrate level phosphorylation in the Krebs cycle. Notice that the Krebs cycle in itself does not produce much ATP, but functions mostly in the transfer of electrons to be used in the electron transport chain that makes the most ATP.

In the **Electron Transport Chain,** NADH transfers electrons from glycolysis and the Kreb's cycle to the first molecule in the chain of molecules embedded in the inner membrane of the mitochondrion.

Most of the molecules in the electron transport chain are proteins. Nonprotein molecules are also part of the chain and are essential for the catalytic functions of certain enzymes. The electron transport chain does not make ATP directly. Instead, it breaks up a large free energy drop into a more manageable one. The chain uses electrons to pump H^+ ions across the mitochondrion membrane.

The H^+ gradient is used to form ATP in a process called **chemiosmosis** (oxidative phosphorylation). ATP synthetase and energy generated by the movement of hydrogen ions coming off of NADH and $FADH_2$ builds ATP from ADP on the inner membrane of the mitochondria. Each NADH yields three molecules of ATP (10 x 3) and each $FADH_2$ yields two molecules of ATP (2 x 2). Thus, the electron transport chain and oxidative phosphorylation produces 34 ATP.

Thus, the net gain from the whole process of respiration is 36 molecules of ATP:

Process	# ATP produced (+)	# ATP consumed (-)	Net # ATP
Glycolysis	4	2	+2
Acetyl CoA	0	2	-2
Krebs cycle	1 per cycle (2 cycles)	0	+2
Electron transport chain	34	0	+34
Total			+36

Photosynthesis is an anabolic process that stores energy in the form of a three carbon sugar. We will use glucose as an example for this section.

Photosynthesis occurs only in organisms that contain chloroplasts (i.e., plants, some bacteria, and some protists). There are a few terms to be familiar with when discussing photosynthesis.

An **autotroph** (self-feeder) is an organism that makes its own food from the energy of the sun or other elements. Autotrophs include:

1. **photoautotrophs** - make food from light and carbon dioxide releasing oxygen that can be used for respiration.

2. **chemoautotrophs** - oxidize sulfur and ammonia. Some bacteria are chemoautotrophs.

Heterotrophs (other feeder) are organisms that must eat other living things to obtain energy. Another term for heterotrophs is **consumers**. All animals are heterotrophs. **Decomposers** break down once living things. Bacteria and fungi are examples of decomposers. **Scavengers** eat dead things. Examples of scavengers are bacteria, fungi, and some animals.

The **chloroplast** is the site of photosynthesis. It is similar to the mitochondria due to the increased surface area of the thylakoid membrane. It also contains a fluid called stroma between the stacks of thylakoids. The thylakoid membrane contains pigments (chlorophyll) that are capable of capturing light energy.

Photosynthesis reverses the electron flow. Water is split by the chloroplast into hydrogen and oxygen. The oxygen is given off as a waste product as carbon dioxide is reduced to sugar (glucose). This requires the input of energy, which comes from the sun.

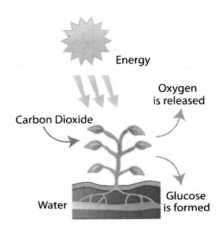

Photosynthesis occurs in two stages, the light reactions and the Calvin cycle (dark reactions). The conversion of solar energy to chemical energy occurs in the light reactions. Electrons are transferred by the absorption of light by chlorophyll and cause the water to split, releasing oxygen as a waste product.

The chemical energy created in the light reaction is in the form of NADPH. ATP is also produced by a process called photophosphorylation. These forms of energy are produced in the thylakoids and are used in the Calvin cycle to produce sugar.

The second stage of photosynthesis is the **Calvin cycle**. Carbon dioxide in the air is incorporated into organic molecules already in the chloroplast. The NADPH produced in the light reaction is used as reducing power for the reduction of the carbon to carbohydrate. ATP from the light reaction is also needed to convert carbon dioxide to carbohydrate (sugar).

The two stages of photosynthesis are summarized below.

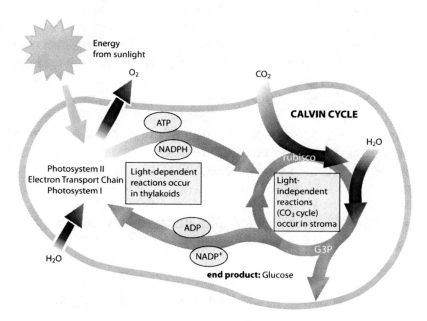

The process of photosynthesis is made possible by the presence of the sun. Visible light ranges in wavelengths of 750 nanometers (red light) to 380 nanometers (violet light). As wavelength decreases, the amount of available energy increases. Light is carried as photons, which are fixed quantities of energy. Light is reflected (what we see), transmitted, or absorbed (what the plant uses). The plant's pigments capture light of specific wavelengths. Remember that the light that is reflected is what we see as color. Plant pigments include:

Chlorophyll *a* - reflects green/blue light; absorbs red light
Chlorophyll *b* - reflects yellow/green light; absorbs red light
Carotenoids - reflects yellow/orange; absorbs violet/blue light

The pigments absorb photons. The energy from the light excites electrons in the chlorophyll that jump to orbitals with more potential energy and reach an "excited" or unstable state.

The formula for photosynthesis is:

$$CO_2 + H_2O + \text{energy (from sunlight)} \rightarrow C_6H_{12}O_6 + O_2$$

The high energy electrons are trapped by primary electron acceptors which are located on the thylakoid membrane. These electron acceptors and the pigments form reaction centers called photosystems that are capable of capturing light energy. Photosystems contain a reaction-center chlorophyll that releases an electron to the primary electron acceptor. This transfer is the first step of the light reactions. There are two photosystems, named according to their date of discovery, not their order of occurrence.

Photosystem I is composed of a pair of chlorophyll *a* molecules. Photosystem I is also called P700 because it absorbs light of 700 nanometers. Photosystem I makes ATP whose energy is needed to build glucose.

Photosystem II is also called P680 because it absorbs light of 680 nanometers. Photosystem II produces ATP + $NADPH_2$ and the waste gas oxygen.

Both photosystems are bound to the **thylakoid membrane**, close to the electron acceptors.

The production of ATP is termed **photophosphorylation** due to the use of light. Photosystem I uses cyclic photophosphorylation because the pathway occurs in a cycle. It can also use noncyclic photophosphorylation which starts with light and ends with glucose. Photosystem II uses noncyclic photophosphorylation only.

Below is a diagram of the relationship between cellular respiration and photosynthesis.

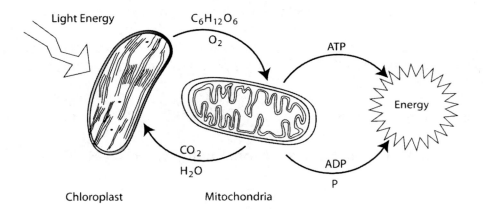

Skill 9.3   Demonstrating an understanding of ATP production through chemiosmosis in both photosynthesis and respiration

Chemiosmosis is the coupling of the diffusion of molecules across a membrane to adenosine triphosphate (ATP) synthesis. In both photosynthesis and respiration, cells use the energy from the transfer of electrons to pump hydrogen ions across the membranes of mitochondria and chloroplasts. This process establishes an electrochemical concentration gradient, where the concentration of hydrogen ions outside of the cellular organelles is much higher than the concentration within the cell. Thus, diffusion of ions back across the membrane yields energy.

Mitochondria and chloroplasts use this energy to create the high-energy molecule ATP by adding a phosphate group to adenosine diphosphate (ADP).

Cellular respiration is the complete breakdown of glucose to produce energy. The final steps of the respiration process occur in the mitochondria. Cells create the electrochemical concentration gradient across the mitochondrial membrane, using the energy from the transport of electrons to pump ions out of the mitochondria. Hydrogen ions diffuse back into mitochondria, toward the area of lower concentration, through an enzyme called ATP synthase. This energy yielding process drives the phosphorylation of ADP to form ATP.

During photosynthesis, light energy creates a high-energy electron donor. The electron donors pass electrons to electron acceptors, generating energy in the process. As in cellular respiration, plants use the energy produced from the electron transport to pump hydrogen ions across the thylakoid membranes of the chloroplasts. This pumping process creates an electrochemical gradient. Diffusion of ions back across the thylakoid membrane through ATP synthase yields energy. Plants use the energy to produce high-energy ATP by the phosphorylation of ADP.

Skill 9.4   Analyzing factors that affect photosynthesis and respiration

Several environmental and cellular factors affect the rate of photosynthesis and cellular respiration. The three main environmental factors affecting photosynthesis are light intensity, carbon dioxide concentration, and temperature. In addition, cellular and physical plant characteristics including leaf shape, nitrogen content, and molecular carrier (e.g. NADP, FAD) concentration affect photosynthetic rates. The main factors affecting cellular respiration rates are temperature, oxygen concentration, and molecular carrier concentration.

Environmental factors such as light intensity, carbon dioxide concentration, and temperature greatly affect photosynthesis. First, rates of photosynthesis increase as light intensity increases to a certain plateau level. Too much light can inactivate a plant's photosynthetic system. In addition, the rate of photosynthesis increases as carbon dioxide increases until other factors become limiting. Finally, the rate of photosynthesis increases as temperature increases until the temperature reaches a critical point that begins to damage the plant.

The physical and cellular factors affecting photosynthesis are leaf shape, molecular carrier concentration, and nitrogen content. The shape of a leaf affects the efficiency of light absorption. The concentration of molecular carrier molecules such as NADP and FAD affects the rate and efficiency of electron transport. Finally, nitrogen is a necessary nutrient for proper plant functioning.

The main factors affecting cellular respiration rates are temperature, oxygen content, and molecular carrier concentration.

Cellular processes, including respiration, generally increase as temperature increases to a certain plateau level. In addition, oxygen content affects the rate, type (anaerobic versus aerobic), and ultimate energy yield of respiration. Aerobic respiration, requiring oxygen, yields more energy than anaerobic respiration that proceeds in the absence of adequate levels of oxygen. Finally, as with photosynthesis, the cellular concentration of NADP and FAD affects the rate and efficiency of respiration.

Skill 9.5   Compare aerobic and anaerobic respiration

Glycolysis generates ATP with oxygen (aerobic) or without oxygen (anaerobic). We have already discussed aerobic respiration. Anaerobic respiration occurs through fermentation. In the process of fermentation, ATP is generated by substrate level phosphorylation if enough NAD^+ is present to accept electrons during oxidation. In anaerobic respiration, NAD^+ is regenerated by transferring electrons to pyruvate. There are two common types of fermentation.

In **alcoholic fermentation**, pyruvate is converted to ethanol in two steps. In the first step, carbon dioxide is released from the pyruvate. In the second step, ethanol is produced when acetaldehyde is reduced by NADH. This results in the regeneration of NAD^+ for glycolysis. Alcohol fermentation is carried out by yeast and some bacteria.

In **lactic acid fermentation**, pyruvate is reduced by NADH forming lactate as a waste product. Animal cells and some bacteria that do not use oxygen utilize lactic acid fermentation to make ATP. Lactic acid forms when pyruvic acid accepts hydrogen from NADH. A buildup of lactic acid is what causes muscle soreness following exercise.

Energy remains stored in lactic acid or alcohol until it is needed. This is not an efficient type of respiration. When oxygen is present, aerobic respiration occurs after glycolysis.

Both aerobic and anaerobic pathways oxidize glucose to pyruvate through the process of glycolysis and both pathways employ NAD^+ as an oxidizing agent. A substantial difference between the two pathways is that in fermentation, an organic molecule such as pyruvate or acetaldehyde is the final electron acceptor. In respiration, the final electron acceptor is oxygen. Another key difference is that respiration yields much more energy from a sugar molecule than fermentation does. Respiration can produce up to 18 times more ATP than fermentation.

Skill 9.6 Evaluating the significance of chloroplast structure in photosynthesis and mitochondrion structure in respiration

Within chloroplasts, the process of photosynthesis takes place across a membrane to convert light energy into chemical energy stored in sugar.

In mitochondria, a similar reaction, cellular respiration, also takes place across a membrane to create ATP. The processes are complementary in that photosynthesis allows energy from light to enter the food chain (as sugar) and cellular respiration harnesses that energy (from sugar) into a form usable by all other processes essential to maintaining life (ATP). These two reactions are among the most important in the carbon cycle, the series of biochemical reactions that move carbon from the atmosphere, to living things, into geological structures, and back to the atmosphere.

Chloroplasts and mitochondria are similar in their suspected origins. Most biologists now believe that both these organelles were originally separate prokaryotic organisms that were drawn into cells as endosymbionts (organisms that live inside other organisms for their mutual benefit). This hypothesized origin of these organelles is the endosymbiotic theory. This may explain the similar structures of chloroplasts and mitochondria; both have inner and outer membranes. The outer membrane may correspond to the cell membrane of the original endosymbiont. Additionally, both mitochondria and chloroplasts have their own DNA and it exists in circular form, as it does in prokaryotes. Both these organelles are also able to synthesize proteins using ribosomes similar to those seen in bacteria. Finally, both mitochondria and chloroplasts are able to reproduce themselves independently of the cell as a whole. At present, it is suspected that mitochondria evolved from aerobic bacteria and chloroplasts from cyanobacteria.

There are also similarities in the manner in which these organelles accomplish their unique energy transformation. Both mitochondria and chloroplasts have highly specific trans-membranous proteins that perform the energy transformation. In mitochondria, the inner membrane is the location of all the integral proteins needed in the electron transport chain and ATP synthase. The maintenance of a proton gradient across the inner membrane is critical to the process of ATP synthesis. Chloroplasts have one more membrane bound compartment than mitochondria: the thylakoids. The thylakoid membrane contains the proteins of the electron transport change and the chlorophyll-containing photosystems that are excited by light. The proton gradient is established across this membrane, rather than across the inner membrane, as happens in cellular respiration. However, photosynthesis, just like cellular respiration, relies on cross-membrane fluxes to drive biosynthesis.

COMPETENCY 10.0   UNDERSTAND THE STRUCTURE AND FUNCTION OF DNA AND RNA

Skill 10.1   Demonstrating an understanding of the mechanism of DNA replication, potential errors, and implications of these errors

DNA replicates semi-conservatively meaning the two original DNA strands are conserved and serve as a template for the new strand.

In DNA replication, the first step is to separate the two strands. As they separate, they unwind the supercoils to reduce tension. An enzyme called **helicase** unwinds the DNA as the replication fork proceeds and **topoisomerases** relieve the tension by nicking one strand and relaxing the supercoil.

Once the strands have separated, they must be stabilized. Single-strand binding proteins (SSBs) bind to the single strands until the DNA is replicated.

An RNA polymerase called primase adds ribonucleotides to the DNA template to initiate DNA synthesis. This short RNA-DNA hybrid is called a **primer**. Once the DNA is single stranded, **DNA polymerases** add nucleotides in the 5' → 3' direction.

As DNA synthesis proceeds along the replication fork, it becomes obvious that replication is semi-discontinuous; meaning one strand is synthesized in the direction the replication fork is moving and the other is synthesized in the opposite direction. The continuously synthesized strand is the **leading strand** and the discontinuously synthesized strand is the **lagging strand**. As the replication fork proceeds, new primer is added to the lagging strand and it is synthesized discontinuously in small fragments called **Okazaki fragments**.

The remaining RNA primers must be removed and replaced with deoxyribonucleotides. DNA polymerase has 5' → 3' polymerase activity and has 3' → 5' exonuclease activity. This enzyme binds to the nick between the Okazaki fragment and the RNA primer. It removes the primer and adds deoxyribonucleotides in the 5' → 3' direction. The nick remains until **DNA ligase** seals it, producing the final product, a double-stranded segment of DNA.

Once the double-stranded segment is replicated, there is a proofreading system carried out by DNA replication enzymes. In eukaryotes, DNA polymerases have 3' → 5' exonuclease activity—they move backwards and remove nucleotides when the enzyme recognizes an error, then add the correct nucleotide in the 5' → 3' direction. In bacteria, DNA polymerase III is the main polymerase that elongates DNA during replication and has exonuclease proofreading ability.

DNA Replication

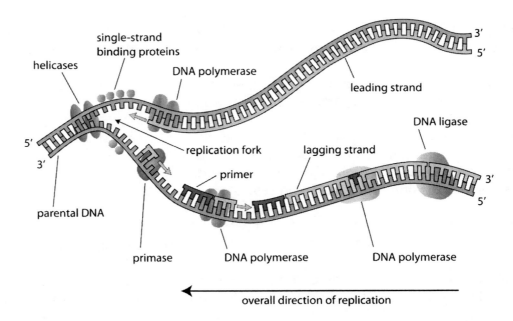

Chromosomal replication in bacteria is similar to eukaryotic DNA replication.

A **plasmid** is a small ring of DNA that carries accessory genes separate from those of the bacterial chromosome. Most plasmids in Gram-negative bacteria undergo bidirectional replication, although some replicate unidirectionally because of their small size. Plasmids in Gram-positive bacteria replicate by the rolling circle mechanism.

Some plasmids can transfer themselves (and therefore their genetic information) through a process called conjugation. Conjugation requires cell-to-cell contact. The sex pilus of the donor cell attaches to the recipient cell. Once contact has been established, the transfer of DNA occurs by the rolling circle mechanism.

Skill 10.2   Analyzing the roles of DNA and ribosomal, messenger, and transfer RNA in protein synthesis

of RNA are needed to carry out these processes: messenger RNA (mRNA), ribosomal RNA (rRNA), and transfer RNA (tRNA). **Messenger RNA** contains information for translation. **Ribosomal RNA** is a structural component of the ribosome and **transfer RNA** carries amino acids to the ribosome for protein synthesis.

Transcription is similar in prokaryotes and eukaryotes. During transcription, the DNA molecule is copied into an RNA molecule (mRNA). Transcription occurs through the steps of initiation, elongation, and termination. Transcription also occurs for rRNA and tRNA, but the focus here is on mRNA.

Initiation begins at the promoter of the double-stranded DNA molecule. The promoter is a specific region of DNA that directs the **RNA polymerase** to bind to the DNA. The double-stranded DNA opens up and RNA polymerase begins transcription in the 5' → 3' direction by pairing ribonucleotides to the deoxyribonucleotides as follows to get a complementary mRNA segment:

Deoxyribonucleotide		Ribonucleotide
A	→	U
G	→	C

Elongation is the synthesis of the mRNA strand in the 5' → 3' direction. The new mRNA rapidly separates from the DNA template and the complementary DNA strands pair together.

Termination of transcription occurs at the end of a gene. Cleavage occurs at specific sites on the mRNA. This process is aided by termination factors.

In eukaryotes, mRNA goes through **posttranscriptional processing** before going on to translation.

There are three basic steps of processing:

1. **5' capping** – The addition of a base with a methyl attached to it that protects the 5' end from degradation and serves as the site where ribosomes bind to the mRNA for translation.

2. **3' polyadenylation** – The addition of 100-300 adenines to the free 3' end of mRNA resulting in a poly-A-tail.

3. **Intron removal**- The removal of non-coding introns and the splicing together of coding exons form the mature mRNA.

Translation is the process in which the mRNA sequence becomes a polypeptide. The mRNA sequence determines the amino acid sequence of a protein by following a pattern called the genetic code. The **genetic code** consists of 64 triplet nucleotide combinations called **codons**. Three codons are termination codons and the remaining 61 code for amino acids. There are 20 amino acids encoded by different mRNA codons. Amino acids are the building blocks of protein. They are attached together by peptide bonds to form a polypeptide chain.

Ribosomes are the site of translation. They contain rRNA and many proteins. Translation occurs in three steps: initiation, elongation, and termination. Initiation occurs when the methylated tRNA binds to the ribosome to form a complex. This complex then binds to the 5' cap of the mRNA. In elongation, tRNAs carry the amino acid to the ribosome and place it in order according to the mRNA sequence. tRNA is very specific – it only accepts one of the 20 amino acids that correspond to the anticodon. The anticodon is complementary to the codon. For example, using the codon sequence below:

the mRNA reads   A U G / G A G / C A U / G C U
the anticodons are U A C / C U C / G U A / C G A

Termination occurs when the ribosome reaches any one of the three stop codons: UAA, UAG, or UGA. The newly formed polypeptide then undergoes posttranslational modification to alter or remove portions of the polypeptide.

Skill 10.3   Analyzing the implications of mutations in DNA molecules for protein structure and function (e.g., sickle-cell anemia, cystic fibrosis)

Inheritable changes in DNA are called mutations. **Mutations** may be errors in replication or a spontaneous rearrangement of one or more segments by factors like radioactivity, drugs, or chemicals. The severity of the change is not as critical as where the change occurs. DNA contains large segments of non-coding areas called introns. The important coding areas are called exons. If an error occurs on an intron, there is no effect. If the error occurs on an exon, it may be minor to lethal depending on the severity of the mistake. Mutations may occur on somatic or sex cells. Usually the mutations on sex cells are more dangerous since they contain the basis of all information for the developing offspring. But mutations are not always bad. They are the basis of evolution and if they create a favorable variation that enhances the organism's survival they are beneficial. But mutations may also lead to abnormalities, birth defects, and even death. There are several types of mutations.

A **point mutation** is a mutation involving a single nucleotide or a few adjacent nucleotides. Let's suppose a normal sequence was as follows:

Normal sequence	A B C D E F
Duplication (a nucleotide is repeated)	A B **C C** D E F
Inversion (a segment is reversed)	A **E D C B** F
Insertion or **Translocation** (a segment of DNA is put in the wrong location)	A B C **R S** D E F
Deletion (a segment is lost)	A B C (DEF lost)

Deletion and insertion mutations that shift the reading frame are **frame shift mutations**. A **silent mutation** makes no change in the amino acid sequence, therefore it does not alter the protein function. A **missense mutation** results in an alteration in the amino acid sequence. A mutation's effect on protein function depends on which amino acids are involved and how many are involved. The structure of a protein usually determines its function. A mutation that does not alter the structure will probably have little or no effect on the protein's function. However, a mutation that does alter the structure of a protein and can severely affect protein activity is called a **loss-of-function mutation**. Sickle-cell anemia and cystic fibrosis are examples of loss-of-function mutations.

Sickle-cell anemia is characterized by weakness, heart failure, joint and muscular impairment, fatigue, abdominal pain and dysfunction, impaired mental function, and eventual death. The mutation that causes this genetic disorder is a point mutation in the sixth amino acid of hemoglobin. A normal hemoglobin molecule has glutamic acid as the sixth amino acid and the sickle-cell hemoglobin has valine at the sixth position. This mutation causes the chemical properties of hemoglobin to change. The hemoglobin of a sickle-cell person has a lower affinity for oxygen, causing red blood cells to have a sickle shape. The sickle shape of the red blood cell causes the formation of clogs because the cells do not pass through capillaries well.

Cystic fibrosis is the most common genetic disorder of people with European ancestry. This disorder affects the exocrine system. A fibrous cyst is formed on the pancreas that blocks the pancreatic ducts. This causes sweat glands to release high levels of salt. A thick mucous is secreted from mucous glands, which accumulates in the lungs. This accumulation of mucous causes bacterial infections and possible death. Cystic fibrosis cannot be cured, but can be treated for a short while. Most children with the disorder die before adulthood. Scientists identified a protein that transports chloride ions across cell membranes. Those with cystic fibrosis have a mutation in the gene coding for the protein. The majority of the mutant alleles have a deletion of the three nucleotides coding for phenylalanine at position 508. Other people with the disorder have mutant alleles caused by substitution, deletion, and frameshift mutations.

Skill 10.4   Analyzing the control of gene expression in cells (e.g., lac operon in *E. coli*)

In bacterial cells, the *lac* operon is a good example of the control of gene expression. The *lac* operon contains the genes that code for the enzymes used to convert lactose into fuel (glucose and galactose). The *lac* operon contains three genes, *lac Z*, *lac Y*, and *lac A*. *Lac Z* codes for an enzyme that converts lactose into glucose and galactose. *Lac Y* codes for an enzyme that causes lactose to enter the cell. *Lac A* codes for an enzyme that acetylates lactose.

The *lac* operon also contains a promoter and an operator that is the "off and on" switch for the operon. A protein called the repressor switches the operon off when it binds to the operator. When lactose is absent, the repressor is active and the operon is turned off. The operon is turned on again when allolactose (formed from lactose) inactivates the repressor by binding to it.

COMPETENCY 11.0   UNDERSTAND THE PROCEDURES INVOLVED IN THE ISOLATION, MANIPULATION, AND EXPRESSION OF GENETIC MATERIAL AND THE APPLICATION OF GENETIC ENGINEERING IN BASIC AND APPLIED RESEARCH

Skill 11.1   Analyzing the role and applications of genetic engineering in the basic discoveries of molecular genetics (e.g., in medicine, agriculture)

Genetic engineering has made enormous contributions to medicine and has opened the door to DNA technology.

The use of DNA probes and the polymerase chain reaction (PCR) has enabled scientists to identify and detect elusive pathogens. Diagnosis of genetic disease is now possible before the onset of symptoms.

Genetic engineering has allowed for the treatment of some genetic disorders. **Gene therapy** is the introduction of a normal allele to the somatic cells to replace the defective allele. The medical field has had success in treating patients with a single enzyme deficiency disease. Gene therapy has allowed doctors and scientists to introduce a normal allele that would provide the missing enzyme.

Insulin and mammalian growth hormones have been produced in bacteria by gene-splicing techniques. Insulin treatment helps control diabetes for millions of people who suffer from the disease. The insulin produced in genetically engineered bacteria is chemically identical to that made in the pancreas. Human growth hormone (HGH) has been genetically engineered for treatment of dwarfism caused by insufficient amounts of HGH. HGH is being further researched for treatment of broken bones and severe burns.

Biotechnology has advanced the techniques used to create vaccines. Genetic engineering allows for the modification of a pathogen in order to attenuate it for vaccine use. In fact, vaccines created by a pathogen attenuated by gene-splicing may be safer than those that use the traditional mutants.

Forensic scientists regularly use DNA technology to solve crimes. DNA testing can determine a person's guilt or innocence. A suspect's DNA fingerprint is compared to the DNA found at the crime scene. If the fingerprints match, guilt can then be established.

Biotechnology has benefited agriculture also. For example, many dairy cows are given bovine growth hormone to increase milk production. Commercially grown plants are often genetically modified for optimal growth.

Strains of wheat, cotton, and soybeans have been developed to resist herbicides used to control weeds. This allows for the successful growth of the plants while destroying the weeds.

Crop plants are also being engineered to resist infections and pests. Scientists can genetically modify crops to contain a viral gene that does not affect the plant and will "vaccinate" the plant from a virus attack. Crop plants are now being modified to resist insect attacks. This allows for farmers to reduce the amount of pesticide used on plants.

Skill 11.2 Demonstrating an understanding of genetic engineering techniques (e.g., restriction enzymes, PCR, gel electrophoresis)

In its simplest form, genetic engineering requires enzymes to cut DNA, a vector, and a host organism for the recombinant DNA. A **restriction enzyme** is a bacterial enzyme that cuts foreign DNA in specific locations. The restriction fragment that results can be inserted into a bacterial plasmid (**vector**). Other vectors that may be used include viruses and bacteriophages. The splicing of restriction fragments into a plasmid results in a recombinant plasmid. This recombinant plasmid can then be placed in a host cell, usually a bacterial cell, for replication.

The use of recombinant DNA provides a means to transplant genes among species. This opens the door for cloning specific genes of interest. Hybridization can be used to find a gene of interest. A probe is a molecule complementary in sequence to the gene of interest. The probe, once it has bonded to the gene, can be detected by labeling with a radioactive isotope or a fluorescent tag.

One of the most widely used genetic engineering techniques is the **polymerase chain reaction (PCR)**. PCR is a technique in which a piece of DNA can be amplified into billions of copies within a few hours. This process requires a primer to specify the segment to be copied, and an enzyme (usually taq polymerase) to amplify the DNA. PCR has allowed scientists to perform multiple procedures on small amounts of DNA.

Gel electrophoresis is another method for analyzing DNA. Electrophoresis separates DNA or protein by size or electrical charge. The DNA runs towards the positive charge and the DNA fragments separate by size. The gel is treated with a DNA-binding dye that fluoresces under ultraviolet light. Gels are typically run with multiple lanes (here there are five) where a different sample is run in each lane and at least one lane is a control as required by the scientific method. . A picture of the gel can be taken and used for analysis. The result looks like the following picture.

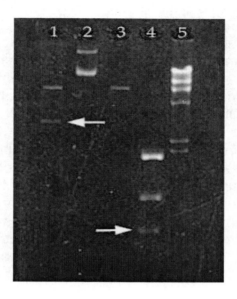

Photograph of gel

Skill 11.3   Analyzing the role of genetic engineering in the development of microbial cultures capable of producing valuable products (e.g., human insulin, growth hormone)

The medical field has had success in treating patients with a single enzyme deficiency disease. Gene therapy has allowed doctors and scientists to introduce a normal allele that provides the missing enzyme.

Insulin and mammalian growth hormones have been produced in bacteria by gene-splicing techniques. Insulin treatment helps control diabetes for millions of people who suffer from the disease. The insulin produced in genetically engineered bacteria is chemically identical to that made in the pancreas. Human growth hormone (HGH) has been genetically engineered for treatment of dwarfism caused by insufficient amounts of HGH. HGH is being further researched for treatment of broken bones and severe burns.

Skill 11.4   Recognizing the role of gene cloning in deriving nucleotide and amino acid sequences and the role of cloned genes as probes in determining the structure of more complex DNA molecules

DNA sequencing has become an area of high interest within the science community during the last century. Nucleic acid research is both exciting and challenging. Investigating a complex DNA sample with unknown nucleotide sequences requires the use of in vivo cloning. In vivo cloning has become a very useful and reliable technology in this type of research. Cloning in bacteria, in phages and in other in vivo systems, is the only way to make the sample suitable for screening and sequencing.

PCR (Polymerase Chain Reaction) is a molecular biological technique for enzymatically replicating DNA without using any living organism such as *E.coli* or yeast. PCR amplifies a small amount of DNA exponentially. This technique is widely used in experiments involving gene manipulations. PCR is also of use in paternity testing and DNA fingerprinting.

The main task of cloning is the amplification of individual molecules. The cloning that we describe here will work for any random piece of DNA. The goal of many experiments is to obtain a sequence of DNA that directs the production of a specific protein. The technique that is most useful for this type of experimentation is cDNA cloning. The principle behind this technique is that an mRNA population isolated from a specific developmental stage should contain mRNAs specific for any protein expressed during that stage. Thus, if we can isolate mRNA, we can study the gene.

There are four main uses of these isolated genes:

1. Deriving the nucleic acid sequence of the gene
2. Studying the sequences of the regulatory region of the gene
3. Isolating similar genes from other organisms
4. Diagnostic purposes, if the gene is of clinical importance

Amino acid sequencing in a protein is another area of research that is very useful and very popular. Amino acid sequencing is determining the sequence of amino acids of a protein's constituent peptides. This sequencing is important in understanding cellular processes and promotes the invention of drugs to target specific metabolic pathways.

Traditionally, amino acid sequencing is done using mass spectrometry and the Edman degradation reaction. Amino acid sequencing is also done indirectly using PCR technique.

Skill 11.5   Recognizing how genetic engineers design new biological products unavailable from natural sources and alter gene products by site-directed mutagenesis (e.g., transgenic plants and animals)

Genetic engineering is a field that is getting lot of attention from scientists these days because of its many practical applications. Designing new biological products and gene alteration are among these applications. These two tasks are achieved by site directed mutagenesis.

Site directed mutagenesis (SDM) is an important tool used to modify DNA sequences in molecular biology and genetic engineering. It is also used in the study of protein structure and function.

There are a number of mutagenesis methods based on PCR (Polymerase Chain Reaction). The simplest of all these is the "Quick Change Site Directed Mutagenesis System" (QCM), developed by Stratagene (LaJolla, CA). In this approach, the mutation is introduced in a single PCR with one pair of complementary primers containing the mutation of interest. While effective, this method is restricted to primer pairs of 25-40 nucleotide bases in length with melting temperature of around 76°C. Research is proving that site directed mutagenesis can be carried out using the Quick Change method.

Some mutations are beneficial to humans. At the same time, there are some risks involved.

Transgenic plants are obtained through genetic engineering, a breeding approach, which uses recombinant DNA techniques to create plants with new characteristics. These are developed for a number of reasons, such as longer shelf life, disease resistance, herbicide resistance, and pest resistance. The first transgenic crop was approved for sale in the US in 1994, the FlavrSavr tomato, which was designed to have a longer shelf life.

There are four general types of transgenic plants:

- Resistant to disease or herbivory (consumption by insects)
- Resistant to specific herbicides
- Resistant to harsh environmental conditions such as extremes of cold, heat, drought, or salt concentration
- Plants that are healthier or are neutraceuticals - producing higher concentrations of specific compounds like lycopene or beta carotene

An emerging group of transgenic plants called pharmacrops aim to use plants to make pharmaceuticals and industrial chemicals.

There are three main ecological risks associated with transgenic plants:

- Escape of transgenic plants from cultivated to uncultivated areas
- Transgenic plants hybridizing with similar wild plants; transfer of some transgenes could occur
- If these new transgene plants become invasive, they will affect biodiversity and thereby the entire ecosystem.

Transgenic animals are created through DNA microinjection. There are two additional techniques used to create transgenic animals - retrovirus mediated transgenesis and embryonic stem (ES) cell mediated gene transfer. Since 1981, there has been rapid development in the use of genetically engineered animals as researchers have found an increasing number of applications.

Uses of transgenic animals in biotechnology:

- In medical research
- In toxicology testing as responsive test animals
- In mammalian developmental genetics
- In molecular biology to investigate genetic changes
- In pharmaceuticals, for drug testing
- In the production of specific proteins.

There are many more uses of transgenic animals. However, the wide use of animals in testing creates ethical problems due to the issue of animal cruelty.

Skill 11.6   Recognizing the ethical, legal, and social implications of genetic engineering

Genetic engineering is a process that uses technology to manipulate an organism's genes. It is directed by science and the needs of humans. Genetic engineering has drastically advanced biotechnology. For example, improving crops, manufacturing human insulin for those individuals who are unable to do so, and the production of new lineages have all been made possible through genetic engineering. The oncomouse used for cancer research was created through the use of technology and has greatly advanced our knowledge. With these advancements come concerns about safety and ethics.

Many safety concerns have been answered by strict government regulations. The FDA, USDA, EPA, and National Institutes of Health are just a few of the government agencies that regulate pharmaceutical, food, and environmental technology advancements.

Several ethical questions arise when discussing biotechnology. Individuals do not seem to object to modifying the genome of fruit flies, but many groups actively denounce genetic research on mammals. Many religious groups question whether embryonic stem cell research should be allowed. Animal rights activists argue that animal testing is inhumane. Ecologists wonder if manipulated genes will eventually blend with the natural population and if they will harm the original population, thus disrupting the balance of our planet. These are just a few examples of the debates surrounding genetic engineering. There are strong arguments for both sides of the issues and sensitivity should be given to all concerned.

COMPETENCY 12.0   UNDERSTAND THE CELL CYCLE, THE STAGES AND END PRODUCTS OF MEIOSIS AND MITOSIS, AND THE ROLE OF CELL DIVISION IN UNICELLULAR AND MULTICELLULAR ORGANISMS

Skill 12.1   Describing general events in the cell cycle and analyzing the significance of these events

The purpose of cell division is to provide growth and repair in body (somatic) cells and to replenish or create sex cells for reproduction. There are two forms of cell division: mitosis and meiosis. **Mitosis** is the division of somatic cells and **meiosis** is the division of sex cells (eggs and sperm).

Mitosis is divided into two parts: the **mitotic (M) phase** and **interphase**. In the mitotic phase, mitosis and cytokinesis divide the nucleus and cytoplasm, respectively. This phase is the shortest phase of the cell cycle. Interphase is the stage where the cell grows and copies the chromosomes in preparation for the mitotic phase. Interphase occurs in three stages of growth: the **G1** (growth) period, when the cell grows and metabolizes, the **S** (synthesis) period, when the cell makes new DNA, and the **G2** (growth) period, when the cell makes new proteins and organelles in preparation for cell division.

The mitotic phase is a continuum of change, although we divide it into five distinct stages: prophase, prometaphase, metaphase, anaphase, and telophase.

During **prophase**, the cell proceeds through the following steps continuously, without stopping. First, the chromatin condenses to become visible chromosomes. Next, the nucleolus disappears and the nuclear membrane breaks apart. Then, mitotic spindles composed of microtubules form that will eventually pull the chromosomes apart. Finally, the cytoskeleton breaks down and the actions of centrioles push the spindles to the poles or opposite ends of the cell.

During **prometaphase**, the nuclear membrane fragments and allows the spindle microtubules to interact with the chromosomes. Kinetochore fibers attach to the chromosomes at the centromere region. **Metaphase** begins when the centrosomes are at opposite ends of the cell. The centromeres of all the chromosomes are aligned with one another.

During **anaphase**, the centromeres split in half and homologous chromosomes separate. The chromosomes are pulled to the poles of the cell, with identical sets at either end. The last stage of mitosis is **telophase**. Here, two nuclei form with a full set of DNA that is identical to the parent cell. The nucleoli become visible and the nuclear membrane reassembles. A cell plate is seen in plant cells and a cleavage furrow forms in animal cells. The cell pinches into two cells. Finally, cytokinesis, or division of the cytoplasm and organelles, occurs.

Below is a diagram of mitosis.

Mitosis

Interphase	nucleus	x number of chromosomes
	chromosomes — centrioles	
Prophase		chromosomes double (2x) and crossover
	spindle fibers	
Prometaphase		nucleus dissolves and microtubules attach to centromeres
Metaphase		chromosomes align at middle of cell
Anaphase		separated chromosomes pull apart
Telophase		microtubules disappear cell division begins
Cytokinesis		2 cells formed each with x chromosomes

Skill 12.2   Interpreting the results of experiments relating to the eukaryotic cell cycle (e.g., cloning, polyploidy, tissue cultures, pharming)

Cloning

Cloning creates a population of identical cells from a single cell. There are different types of cloning and the technology for cloning can be used for many different purposes. There are three main types of cloning.

1. Recombinant DNA technology/DNA cloning/molecular cloning/gene cloning

All these terms refer to the same process. Here, the scientist transfers a fragment of DNA of interest to a self-replicating genetic element such as a bacterial plasmid. The DNA of interest can be propagated in a foreign host cell. This technology is commonly used in molecular biology labs and has been around since the 1970s. Scientists studying a particular gene often use bacterial plasmids to generate multiple copies of the same gene. To clone a gene, a scientist isolates a DNA fragment containing the gene of interest from chromosomal DNA using restriction enzymes and then unites the gene with a plasmid.

2. Reproductive cloning

Reproductive cloning generates an animal that has the same nuclear DNA as another currently or previously existing animal. Reproductive cloning produced "Dolly" the cloned sheep. Reproductive cloning uses the process of SCNT - "somatic cell nuclear transfer". Scientists transfer genetic material from the nucleus of a donor adult cell to an egg whose nucleus has been removed. The reconstructed cell is treated with chemicals or electric current in order to stimulate cell division and once the cloned embryo reaches a suitable stage , it is transferred to the uterus of a female host in order to stimulate cell division. Finally, once the cloned embryo reaches a suitable stage, it is transferred to the uterus of a female host where it continues to develop until birth.

3. Therapeutic cloning

Also called "embryonic cloning", therapeutic cloning is the production of human embryos for use in research. The aim of this research is not to clone human beings, but to harvest stem cells that researchers can use to study human development and to treat disease. Stem cells are very important because they can be used to generate virtually any type of specialized cell in the human body.

In 2001, scientists from Advanced Cell technologies (ACT) cloned the first human embryo for the purpose of advancing therapeutic research. To achieve this feat, scientists collected eggs from women's ovaries and removed the genetic material from the eggs with a very fine needle. Next, they inserted a skin cell inside the enucleated egg to serve as a new nucleus. The egg began to divide after it was stimulated by ionomycin. The success of these experiments was very limited.

Cloning technologies are useful in learning about gene therapy, genetic engineering of organisms, and genome sequencing.

Some of the risks and limitations of cloning are that reproductive cloning is very expensive and highly inefficient. There may be errors while DNA duplicates inside a host cell.

Physicians from the American Medical Association and scientists with the American Association for the Advancement of Science have issued formal statements advising against human reproductive cloning.

Polyploidy

Polyploidy is the process of genome doubling that gives rise to organisms with multiple sets of chromosomes. The term ploidy actually means the number of complete genomes present in a single cell. Polyploid organisms in general contain multiple combinations of the chromosome sets found in the same or closely related diploid species.

Polyploidy can result from spontaneous duplication of the somatic chromosomes or due to a non-separation of homologous chromosomes during meiosis, giving rise to diploid gametes instead of the usual haploid gametes with one set of chromosomes. This condition of polyploidy can be artificially induced by treating the cells with colchicin, a drug known for its property of inhibiting cell division.

Endopolyploidy is a term used when polyploidy is restricted to a particular tissue in an organism. Good examples for this type of polyploidy are the salivary gland cells in Drosophila and the liver cells in humans. When polyploidy occurs spontaneously in an organism, it is called 'autopolyploidy' and when it occurs during a cross between two species, it is called 'allopolyploidy'. Paleopolyploidy is ancient genome duplication that probably characterized all life. Examples of recently confirmed ancient genome duplications include baker's yeast, mustard weed/thale grass, rice, and some animals.

Polyploid types are termed depending on the number of chromosome sets - triploid (three sets), tetraploid (four sets), pentaploid (five sets), hexaploid (six sets) and octaploid (eight sets). Polyploidy in humans is rare. Triploidy is seen in only one in 10,000 live births. Tetraploidy is much rarer than triploidy.

Polyploid crops: These plants in general are more robust and sturdy compared to the regular plants. During breeding, strong and sturdy plants are selected over their weak counterparts.

Triploid crops -   banana, apple
Tetraploid crops -   durum wheat, maize, cabbage, leek, tobacco, peanut
Hexaploid crops -   chrysanthemum, bread wheat, oat, triticale
Octaploid crops -   dahlia, pansies, sugar cane

Pharming

Pharming is defined as the manufacture of medical products from genetically modified plants or animals. Pharming is a combination of "farming" and "pharmaceutical" and refers to the use of genetic engineering to insert genes that code for useful pharmaceutical into host animals or plants that would not otherwise express these genes. As a result, the host animals or plants make the pharmaceutical product in large quantities, which can then be purified and used as drugs.

Some examples of pharming are tomato plants that produce very high amounts of lycopene and cows that produce milk that is high in protein. The production technology is competitive and cost effective and is safe, quick, and efficient.

Skill 12.3 **Comparing chromosomal changes during the stages of meiosis and mitosis**

Meiosis is similar to mitosis, but there are two consecutive cell divisions, meiosis I and meiosis II in order to reduce the chromosome number by one half. This way, when the sperm and egg join during fertilization, the haploid number is reached.

Similar to mitosis, meiosis is preceded by an interphase during which the chromosome replicates. The steps of meiosis are as follows:

1. **Prophase I** – The replicated chromosomes condense and pair with homologues in a process called synapsis. This forms a tetrad. Crossing over, the exchange of genetic material between homologues to further increase diversity, occurs during prophase I.

2. **Metaphase I** – The homologous pairs attach to spindle fibers after lining up in the middle of the cell.

3. **Anaphase I** – The sister chromatids remain joined and move to the poles of the cell.

4. **Telophase I** – The homologous chromosome pairs continue to separate. Each pole now has a haploid chromosome set. Telophase I occurs simultaneously with cytokinesis. In animal cells, a cleavage furrow forms and, in plant cells, a cell plate appears.

5. **Prophase II** – A spindle apparatus forms and the chromosomes condense.

6. **Metaphase II** – Sister chromatids line up in center of cell. The centromeres divide and the sister chromatids begin to separate.

7. **Anaphase II** – The separated chromosomes move to opposite ends of the cell.

8. **Telophase II** – Cytokinesis occurs, resulting in four haploid daughter cells.

The following is a diagram of meiosis.

Meiosis

Interphase	x number of chromosomes
Prophase	chromosomes double (2x) and crossover
Prometaphase	nucleus dissolves and microtubules attach to centromeres
Metaphase I	chromosomes align at middle of cell
Anaphase I	separated chromosomes pull apart
Telophase I	microtubules disappear cell division begins
Prophase II	2 cells formed each with x chromosomes
Metaphase II	microtubules attach to centromeres
Anaphase II	chromosomes pull apart
Telophase II	microtubules disappear cell division begins
Cytokinesis	4 cells form each with half the number of original chromosomes (½ x)

Labels in diagram: nucleus, chromosomes, centrioles, spindle fibers

Skill 12.4 **Analyzing the significance of meiosis and fertilization in relation to the genetic diversity and evolution of multicellular organisms**

Meiosis and fertilization are responsible for genetic diversity. There are several mechanisms that contribute to genetic variation in sexual reproductive organisms. Three of them are independent assortment of chromosomes, crossing over, and random fertilization.

At the metaphase I stage of meiosis, each homologous pair of chromosomes is situated along the metaphase plate. The orientation of the homologous pairs is random and independent of the other pairs of metaphase I. This results in an **independent assortment** of maternal and paternal chromosomes. Based on this information, it seems as though each chromosome in a gamete would be of only maternal or paternal origin. A process called crossing over prevents this from happening.

Crossing over occurs during prophase I. At this point, nonsister chromatids cross and exchange corresponding segments. Crossing over results in the combination of DNA from both parents, allowing for greater genetic variation in sexual life cycles.

Random fertilization also results in genetic variation. Each parent has about 8 million possible chromosome combinations. This allows for over 60 trillion diploid combinations.

Skill 12.5 **Recognizing the relationship between an unrestricted cell cycle and cancer**

The restriction point in the cell cycle occurs late in the G_1 phase. This is when the decision for the cell to divide is made. If all the internal and external cell systems are working properly, the cell proceeds to replicate. Cells may also decide not to proceed past the restriction point. This nondividing cell state is called the G_0 phase. Many specialized cells remain in this state.

The density of cells also regulates cell division. Density-dependent inhibition is when the cells crowd one another and consume all the nutrients, therefore halting cell division. Cancer cells do not respond to density-dependent inhibition. They divide excessively and invade other tissues. As long as there are nutrients, cancer cells are "immortal."

Skill 12.6   Demonstrating an understanding of the process of cell differentiation, including the role of stem cells

Differentiation is the process in which cells become specialized in structure and function. The fate of the cell is usually maintained through many subsequent generations. Gene regulatory proteins can generate many cell types during development.   Scientists believe that these proteins are passed down to the next generation of cells to ensure the specialized expression of the genes occurs.

Stem cells are not terminally differentiated. They can divide for as long as the animal is alive. When the stem cell divides, its daughter cells can either remain a stem cell or proceed with terminal differentiation. There are many types of stem cells that are specialized for different classes of terminally differentiated cells.

Embryonic stem cells give rise to all the tissues and cell types in the body. In culture, these cells have led to the creation of animal tissue that can replace damaged tissues. It is hopeful that with continued research, embryonic stem cells can be cultured to replace damaged muscles, tissues, and organs.

Animal tissue becomes specialized during development. The ectoderm (outer layer) becomes the epidermis or skin. The mesoderm (middle layer) becomes muscles and other organs beside the gut. The endoderm (inner layer) becomes the gut, also called the archenteron. A diagram of this can be seen on the following page.

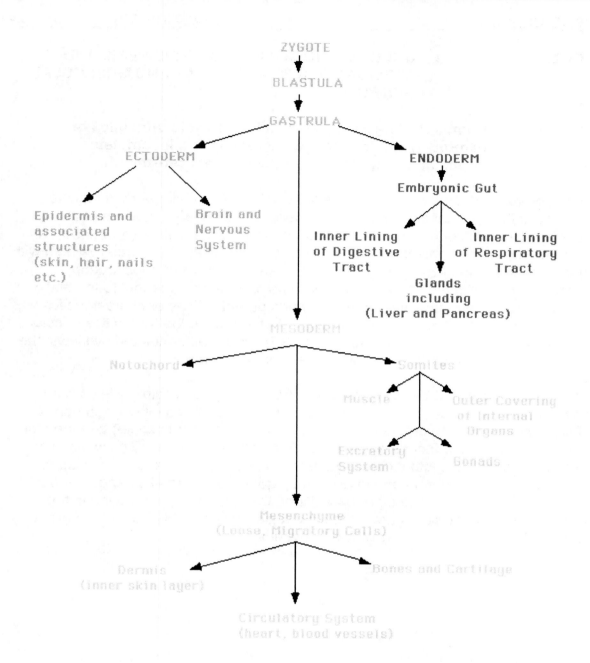

COMPETENCY 13.0   UNDERSTAND CONCEPTS, PRINCIPLES, AND APPLICATIONS OF CLASSICAL AND MOLECULAR GENETICS

Skill 13.1   Demonstrating an understanding of basic principles of heredity (e.g., dominance, codominance, incomplete dominance, segregation, independent assortment)

Gregor Mendel is recognized as the father of genetics. His work in the late 1800's is the basis of our knowledge of genetics. Although unaware of the presence of DNA or genes, Mendel realized there were factors (now known as **genes**) that were transferred from parents to their offspring. Mendel worked with pea plants and fertilized the plants himself, keeping track of subsequent generations which led to the Mendelian laws of genetics. Mendel found that two "factors" governed each trait, one from each parent. Traits or characteristics came in several forms, known as **alleles**. For example, the trait of flower color had white alleles (*pp*) and purple alleles (*PP*). Mendel formulated two laws: the law of segregation and the law of independent assortment.

The **law of segregation** states that only one of the two possible alleles from each parent is passed on to the offspring. If the two alleles differ, then one is fully expressed in the organism's appearance (the dominant allele) and the other has no noticeable effect on appearance (the recessive allele). The two alleles for each trait segregate into different gametes. A Punnet square can be used to show the law of segregation. In a Punnet square, one parent's genes are put at the top of the box and the other parent's on the side. Genes combine in the squares just like numbers are added in addition tables. This Punnet square shows the result of the cross of two F_1 hybrids.

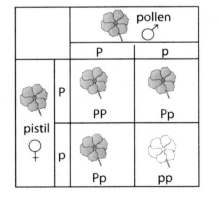

P   PP x pp
              ↓
F_1 Pp x Pp
              ↓
F_2 ¼PP + ½Pp + ¼pp

This cross results in a 1:2:1 ratio of F_2 offspring. Here, the *P* is the dominant allele and the *p* is the recessive allele. The F_1 cross produces three offspring with the dominant allele expressed (two *PP* and *Pp*) and one offspring with the recessive allele expressed (*pp*).

Some other important terms to know:

Homozygous – having a pair of identical alleles. For example, *PP* and *pp* are homozygous pairs.

Heterozygous – having two different alleles. For example, *Pp* is a heterozygous pair.

Phenotype – the organism's physical appearance.

Genotype – the organism's genetic makeup. For example, *PP* and *Pp* have the same phenotype (purple in color), but different genotypes.

The **law of independent assortment** states that alleles assort independently of each other. The law of segregation applies for monohybrid crosses (only one character, in this case flower color, is experimented with). In a dihybrid cross, two characters are explored. Two of the seven characters Mendel studied were seed shape and color. Yellow is the dominant seed color (*Y*) and green is the recessive color (*y*). The dominant seed shape is round (*R*) and the recessive shape is wrinkled (*r*). A cross between a plant with yellow round seeds (*YYRR*) and a plant with green wrinkled seeds (*yyrr*) produces an F_1 generation with the genotype *YyRr*. The production of F_2 offspring results in a 9:3:3:1 phenotypic ratio.

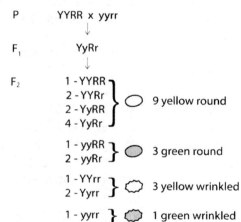

Based on Mendelian genetics, the more complex hereditary pattern of **dominance** was discovered. In Mendel's law of segregation, the F_1 generation has either purple or white flowers. This is an example of **complete dominance**. **Incomplete dominance** is when the F_1 generation results in an appearance somewhere between the two parents. For example, red flowers are crossed with white flowers, resulting in an F_1 generation with pink flowers. The red and white traits are still carried by the F_1 generation, resulting in an F_2 generation with a phenotypic ratio of 1:2:1. In **codominance,** the genes may form new phenotypes. The ABO blood grouping is an example of codominance. A and B are of equal strength and O is recessive. Therefore, type A blood may have the genotypes of AA or AO, type B blood may have the genotypes of BB or BO, type AB blood has the genotype A and B, and type O blood has two recessive O genes.

Skill 13.2　Analyzing techniques used to determine the presence of human genetic diseases (e.g., PKU, cystic fibrosis)

Some genetic disorders can be prevented. The parents can be screened for genetic disorders before the child is conceived or in the early stages of pregnancy. Genetic counselors determine the risk of producing offspring that may express a genetic disorder. The counselor reviews the family's pedigree and determines the frequency of a recessive allele. While genetic counseling is helpful for future parents, there is no certainty in the outcome.

There are some genetic disorders that can be discovered in a heterozygous parent. For example, sickle-cell anemia and cystic fibrosis alleles can be discovered in carriers by genetic testing. If the parents are carriers but decide to have children anyway, fetal testing is available during the pregnancy. There are a few techniques available to determine if a developing fetus will have the genetic disorder.

Amniocentesis is a procedure in which a needle is inserted into the uterus to extract some of the amniotic fluid surrounding the fetus. Some disorders can be detected by chemicals in the fluid. Other disorders can be detected by karyotyping cells cultured from the fluid to identify certain chromosomal defects. In a technique called **chorionic villus sampling (CVS)**, a physician removes some of the fetal tissue from the placenta. The cells are then karyotyped as they are in amniocentesis. The advantage of CVS is that the cells can be karyotyped immediately, unlike in amniocentesis which take several weeks to culture. Unlike amniocentesis and CVS, **ultrasounds** are a non-invasive technique for detecting genetic disorders. Ultrasound can only detect physical abnormalities of the fetus. Newborn screening is now routinely performed in the United States at birth.

Phenylketonuria (PKU) is a recessively inherited disorder that does not allow children to properly break down the amino acid phenylalanine. This amino acid and its by-product accumulate in the blood to toxic levels, resulting in mental retardation. This can be prevented by screening at birth for this defect and treating it with a special diet.

Skill 13.3 Analyzing genetic inheritance problems involving genotypic and phenotypic frequencies

The same techniques of pedigree analysis apply when tracing inherited disorders. Thousands of genetic disorders are the result of inheriting a recessive trait. These disorders range from non-lethal traits (such as albinism) to life-threatening (such as cystic fibrosis).

Most people with recessive disorders are born to parents with normal phenotypes. The mating of heterozygous parents would result in an offspring genotypic ratio of 1:2:1; thus 1 out of 4 offspring would express this recessive trait. The heterozygous parents are called carriers because they do not express the trait phenotypically but can pass the trait on to their offspring.

Lethal dominant alleles are much less common than lethal recessives. This is because lethal dominant alleles are not masked in heterozygotes. Mutations in a gene of the sperm or egg can result in a lethal dominant allele, usually killing the developing offspring.

Sex linked traits - The Y chromosome found only in males (XY) carries very little genetic information, whereas the X chromosome found in females (XX) carries very important information. Since men have no second X chromosome to cover up a recessive gene, the recessive trait is expressed more often in men. Women need the recessive gene on both X chromosomes to show the trait. Examples of sex linked traits include hemophilia and color-blindness.

Sex influenced traits - Traits are influenced by the sex hormones. Male pattern baldness is an example of a sex influenced trait. Testosterone influences the expression of the gene, thus, men are more susceptible to hair loss.

Nondisjunction - During meiosis, chromosomes fail to separate properly. One sex cell may get both chromosomes and another may get none. Depending on the chromosomes involved this may or may not be serious. Offspring end up with either a missing chromosome or an extra chromosome. An example of nondisjunction is Down Syndrome, where three copies chromosome 21 are present.

Chromosome Theory - Introduced by Walter Sutton in the early 1900's. In the late 1800's, the processes of mitosis and meiosis were understood. Sutton saw how this explanation confirmed Mendel's "factors". The chromosome theory basically states that genes are located on chromosomes that undergo independent assortment and segregation.

Skill 13.4   Interpreting pedigree charts

A family pedigree is a collection of a family's history for a particular trait. As you work your way through the pedigree of interest, the Mendelian inheritance theories are applied. In tracing a trait, the generations are mapped in a pedigree chart, similar to a family tree but with the alleles present. In a case where both parent have a particular trait and one of two children also express this trait, then the trait is due to a dominant allele. In contrast, if both parents do not express a trait and one of their children does, that trait is due to a recessive allele. An example is shown below.

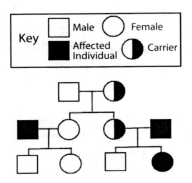

Skill 13.5   Recognizing the role of nonnuclear inheritance (e.g., mitochondrial DNA) in phenotypic expression

Mitochondrial DNA is passed to the next generation exclusively by the mother. A genetic defect in the mother's mitochondrial DNA will always pass to her offspring, regardless of the paternal DNA.

COMPETENCY 14.0   UNDERSTAND THE PRINCIPLES OF POPULATION GENETICS AND THE INTERACTION BETWEEN HEREDITY AND THE ENVIRONMENT, AND APPLY THIS KNOWLEDGE TO PROBLEMS INVOLVING POPULATIONS

Skill 14.1   Evaluating conditions that affect allele frequency in a gene pool

Evolution currently is defined as a change in genotype over time. Gene frequencies shift and change from generation to generation. Populations evolve, not individuals. The **Hardy-Weinberg** theory of gene equilibrium is a mathematical prediction to show shifting gene patterns. Let's use the letter "A" to represent the dominant condition of normal skin pigment, and the letter "a" to represent the recessive condition of albinism. In a population, there are three possible genotypes: *AA, Aa* and *aa*. *AA* and *Aa* would have normal skin pigment and only *aa* would be albinos.

According to the Hardy-Weinberg law, there are five requirements that keep gene frequency stable and limit evolution:

1. There is no mutation in the population.
2. There are no selection pressures; one gene is not more desirable in the environment.
3. There is no mating preference; mating is random.
4. The population is isolated; there is no immigration or emigration.
5. The population is large (mathematical probability is more correct with a large sample).

The above conditions are extremely difficult to meet. If these five conditions are not met, then gene frequency can shift, leading to evolution. Let's say in a population, 75% of the population has normal skin pigment (*AA* and *Aa*) and 25% are albino (*aa*). Using the following formula, we can determine the frequency of the *A* allele and the *a* allele in a population.

This formula can be used over generations to determine if evolution is occurring. The formula is: $1 = p^2 + 2pq + q^2$; where 1 is the total population, p^2 is the number of *AA* individuals, $2pq$ is the number of *Aa* individuals, and q^2 is the number of *aa* individuals.

Since you cannot tell by looking if an individual is *AA* or *Aa*, you must use the *aa* individuals to find that frequency first. As stated above *aa* was 25% of the population. Since $aa = q^2$, we can determine the value of *q* (or *a*) by finding the square root of 0.25, which is 0.5. Therefore, 0.5 of the population has the *a* gene. In order to find the value for *p*, use the following formula: $1 = p + q$. This would make the value of $p = 0.5$.

The gene pool is all the alleles at all gene loci in all individuals of a population. The Hardy-Weinberg theorem describes the gene pool in a non-evolving population. It states that the frequencies of alleles and genotypes in a population's gene pool are random unless acted on by something other than sexual recombination.

Now, to find the number of *AA*, plug it into the first formula:

$$AA = p^2 = 0.5 \times 0.5 = 0.25$$
$$Aa = 2pq = 2(0.5 \times 0.5) = 0.5$$
$$aa = q^2 = 0.5 \times 0.5 = 0.25$$

Any problem you may have with Hardy-Weinberg will have an obvious squared number. The square of that number will be the frequency of the recessive gene, and you can figure anything else out knowing the formula and the frequency of *q*.

When frequencies vary from the Hardy-Weinberg equilibrium, the population is evolving. The change to the gene pool is on such a small scale that it is called microevolution. Certain factors increase the chances of variability in a population, thus leading to evolution. Items that increase variability include mutations, sexual reproduction, immigration, large population, and variation in geographic local. Changes that decrease variation are natural selection, emigration, small population, and random mating.

Skill 14.2   Analyzing the relationship between an organism's phenotype for a particular trait and its selective advantage in a given environment (e.g., the human sickle-cell trait and malaria)

The environment can have an impact on phenotype. For example, a person living at a higher altitude will have a different amount of red and white blood cells than a person living at sea level.

In some cases, a particular trait is advantageous to the organism in a particular environment. Sickle-cell disease causes a low oxygen level in the blood which results in red blood cells having a sickle shape. About one in every ten African-Americans have the sickle-cell trait. These heterozygous carriers are usually healthy compared to homozygous individuals who can suffer severe detrimental effects. In the tropical Africa environment, heterozygotes are more resistant to malaria than those who do not carry any copies of the sickle-cell gene.

COMPETENCY 15.0   UNDERSTAND HYPOTHESES ABOUT THE ORIGINS OF LIFE, EVIDENCE SUPPORTING EVOLUTION, AND EVOLUTION AS A UNIFYING THEME IN BIOLOGY

Skill 15.1   Evaluating evidence supporting various hypotheses about the origins of life

The hypothesis that life developed on Earth from nonliving materials is the most widely accepted theory. The transformation from nonliving materials to life had four stages. The first stage was the nonliving (abiotic) synthesis of small monomers such as amino acids and nucleotides. In the second stage, these monomers combine to form polymers, such as proteins and nucleic acids. The third stage was the accumulation of these polymers into droplets called protobionts. The last stage was the origin of heredity, with RNA as the first genetic material.

The first stage of this theory was hypothesized in the 1920s. A. I. Oparin and J. B. S. Haldane were the first to theorize that the primitive atmosphere was a reducing atmosphere with no oxygen present. The gases were rich in hydrogen, methane, water, and ammonia. In the 1950s, Stanley Miller proved Oparin's theory in the laboratory by combining the above gases. When given an electrical spark, he was able to synthesize simple amino acids. It is commonly accepted that amino acids appeared before DNA. Other laboratory experiments have supported that the other stages in the origin of life theory could have happened.

Other scientists believe simpler hereditary systems originated before nucleic acids. In 1991, Julius Rebek was able to synthesize a simple organic molecule that replicates itself. According to his theory, this simple molecule may be the precursor of RNA.

Skill 15.2   Analyzing the progression from simpler to more complex life forms (e.g., unicellular to colonial to multicellular) by various processes (e.g., endosymbiosis)

Prokaryotes are the simplest life form. Their small genome size limits the number of genes that control metabolic activities. Over time, some prokaryotic groups became multicellular organisms for this reason. Prokaryotes then evolved to form complex bacterial communities where species benefit from one another.

The **endosymbiotic theory** of the origin of eukaryotes states that eukaryotes arose from symbiotic groups of prokaryotic cells. According to this theory, smaller prokaryotes lived within larger prokaryotic cells, eventually evolving into chloroplasts and mitochondria.

Chloroplasts are the descendant of photosynthetic prokaryotes and mitochondria are likely the descendants of bacteria that were aerobic heterotrophs. Serial endosymbiosis is a sequence of endosymbiotic events. Serial endosymbiosis may also play a role in the progression of life forms to become eukaryotes.

Skill 15.3   Assessing the significance of geological and fossil records in determining evolutionary histories and relationships of given organisms

Fossils are the key to understanding biological history. They are the preserved remnants left by an organism that lived in the past. Scientists have established the geological time scale to determine the age of a fossil. The geological time scale is broken down into four eras: the Precambrian, Paleozoic, Mesozoic, and Cenozoic. The eras are further broken down into periods that represent a distinct age in the history of the Earth and its life. Scientists use rock layers called strata to date fossils. The older layers of rock are at the bottom. This allows scientists to correlate the rock layers with the era they date back to. Radiometric dating is a more precise method of dating fossils. Rocks and fossils contain isotopes of elements accumulated over time. The isotope's half-life is used to date older fossils by determining the amount of isotope remaining and comparing it to the half-life.

Dating fossils is helpful in the construction of evolutionary trees. Scientists can arrange the succession of animals based on their fossil record. The fossils of an animal's ancestors can be dated and placed on its evolutionary tree. For example, the branched evolution of horses shows that the modern horse's ancestors were larger, had a reduced number of toes, and had teeth modified for grazing.

For your reference a geologic time table is provided, courtesy of USGS, on the next page.

EON	ERA	PERIOD		EPOCH		Ma
Phanerozoic	Cenozoic	Quaternary		Holocene		
				Pleistocene	Late	0.01
					Early	0.8
		Tertiary	Neogene	Pliocene	Late	1.8
					Early	3.6
				Miocene	Late	5.3
					Middle	11.2
					Early	16.4
				Oligocene	Late	33.7
					Early	28.5
			Paleogene	Eocene	Late	33.7
					Middle	41.3
					Early	49.0
				Paleocene		54.8
					Late	61.0
					Early	65.0
	Mesozoic	Cretaceous		Late		99.0
				Early		144
		Jurassic		Late		159
				Middle		180
				Early		206
		Triassic		Late		227
				Middle		242
				Early		248
	Paleozoic	Permian		Late		256
				Early		290
		Pennsylvanian				323
		Mississippian				354
		Devonian		Late		370
				Middle		391
				Early		417
		Silurian		Late		423
				Early		443
		Ordovician		Late		458
				Middle		470
				Early		490
		Cambrian		D		500
				C		512
				B		520
				A		543
Precambrian	Proterozoic	Late				900
		Middle				1600
		Early				2500
	Archean	Late				3000
		Middle				3400
		Early				3800?

Skill 15.4   Evaluating observations made in various areas of biology (e.g., embryology, biochemistry, anatomy) in terms of evolution

1. Embryology

Comparative embryology shows that embryos start off looking the same. As they develop, their similarities slowly decrease until they take the form of their particular class.

For example, adult vertebrates are diverse, yet their embryos are quite similar at very early stages.

Fish like structures still form in early embryos of reptiles, birds, and mammals. In fish embryos, a two-chambered heart, some veins, and parts of arteries develop and persist in adult fish. The same structures form early in human embryos but do not persist as such in adults.

2. Biochemistry

All known extant organisms make use of DNA and /or RNA. All extant life uses ATP as the metabolic currency. The genetic code is the same for all organisms, meaning that a piece of RNA in a bacterium codes for the same protein as in a human cell.

A classic example of biochemical evidence for evolution is the variance of the protein Cytochrome c in living cells. The variance of Cytochrome c of different organisms is measured in the number of differing amino acids, each differing amino acid being a result of a base pair substitution (i.e. a mutation). If we assume that each differing amino acid is the result of one base pair substitution, we can calculate how long ago the two species diverged by multiplying the number of base pair substitutions by the estimated time it takes for a substituted base pair of the Cytochrome c gene to mutate in N thousand years. The number of amino acids making up the Cytochrome c protein in monkeys differs by one from that of humans. This leads us to believe that the two species diverged N million years ago.

3. Anatomy

Comparative study of the anatomy of the groups of animals or plants reveals that certain structural features are similar. For example, the basic structure of all flowers sepals, petals, stigma, style, and ovary are similar. However, the size, color, number of parts, and specific structure are different for each individual species. The degree of resemblance between to organisms generally indicates how closely related they are in evolution.

Groups with little in common are assumed to have diverged from a common ancestor much earlier in geological history than groups which have a lot in common.

In deciding how closely related two animals are, a comparative anatomist looks for structures which, though they may serve quite different functions in the adult, are fundamentally similar, suggesting a common origin. Such structures are described as homologous; and in cases where the similar structures serve different functions in adults, it may be necessary to trace their origin and embryonic development to look for more similarities derived from a common ancestor.

COMPETENCY 16.0   UNDERSTAND THE MECHANISMS OF EVOLUTION

Skill 16.1   Recognizing sources of variation in a population

Heritable variation is responsible for the individuality of organisms. An individual's phenotype is based on inherited genotype and the surrounding environment.

Mutation and sexual recombination creates genetic variation. **Mutations** may be errors in replication or spontaneous rearrangements of one or more segments of DNA.

Mutations contribute a minimal amount of variation in a population. It is the unique **recombination** of existing alleles that causes the majority of genetic differences. Recombination is caused by the crossing over of the parent genes during meiosis. This results in unique offspring. With all the possible mating combinations in the world, it is obvious how sexual reproduction is the primary cause of genetic variation.

Skill 16.2   Analyzing relationships between changes in allele frequencies and evolution

Natural selection is based on the survival of certain traits in a population through the course of time.   The phrase "survival of the fittest," is often associated with natural selection. Fitness is the contribution an individual makes to the gene pool of the next generation.

Natural selection acts on phenotypes. An organism's phenotype is constantly exposed to its environment. Based on an organism's phenotype, selection indirectly adapts a population to its environment by maintaining favorable genotypes in the gene pool.

There are three modes of natural selection. **Stabilizing selection** favors the more common phenotypes, **directional selection** shifts the frequency of phenotypes in one direction, and **diversifying selection** favors individuals on both extremes of the phenotypic range.

Sexual selection leads to the secondary sex characteristics of males and females. Animals that use mating behaviors may be successful or unsuccessful. A male animal that lacks attractive plumage or has a weak mating call will not attract females, thereby eventually limiting that gene in the gene pool. Mechanical isolation, where sex organs do not fit the female, has an obvious disadvantage.

Skill 16.3   Analyzing the implications of natural selection versus the inheritance of acquired traits in given situations

Natural selection affects populations slowly over prolonged periods. Charles Darwin proposed a mechanism for his theory of evolution, which he termed natural selection. Natural selection describes the process by which favorable traits accumulate in a population, changing the population's genetic make-up over time. Darwin theorized that all individual organisms, even those of the same species, are different and those individuals that happen to possess traits favorable for survival would produce more offspring. Thus, in the next generation, the number of individuals with the favorable trait increases and the process continues.

The inheritance of acquired traits gives an individual a better chance of surviving and reproducing within its own lifespan. For example, lets look at birds. Birds often use plumage to attract a mate. Natural selection has acted upon the gene pool for many generations. If red is an attractive plumage color, then the most red bird is more likely to attract a mate and reproduce, resulting in many offspring with red plumage. While this has been advantageous for many generations, what would happen if the environment were to change? What if a chance mutation created an individual with a color perceived as more attractive to possible mates? Then that new colored individual bird would become more wanted, and then population would start to change, but it would take many generations to see the effect.

The predator-prey relationship also affects population dynamics. If the birds' natural predator suddenly increases in prevalence, the birds will benefit more from camouflage than attractive coloring. Now the bright red mating colors serve as a disadvantage, making the bird more susceptible to attack. What was once a favorable trait is now unfavorable. Based upon this example one can see that an individual's inherited, acquired traits are unique, and may or may not serve the individual well within it's environment. Over time, however, natural selection will create a community of individuals with favorable traits.

Skill 16.4   Comparing alternative mechanisms of evolution (e.g., gradualism, punctuated equilibrium)

There are two theories on the rate of evolution. **Gradualism** is the theory that minor evolutionary changes occur at a regular rate. Darwin's book "On the Origin of Species" is based on this theory of gradualism.

Charles Darwin was born in 1809 and spent 5 years in his twenties on a ship called the *Beagle*. Of all the locations the *Beagle* sailed to, it was the Galapagos Islands that infatuated Darwin. There he collected 13 species of finches that were quite similar. He could not accurately determine whether these finches were of the same species. He later learned these finches were in fact separate species. Darwin began to hypothesize that a new species arose from its ancestors by the gradual collection of adaptations to a different environment. Darwin's most popular hypothesis involves the beak size of Galapagos finches. He theorized that the finches' beak sizes evolved to accommodate different food sources. Many people did not believe in Darwin's theories until recent field studies proved successful.

Although Darwin believed the origin of species was gradual, he was bewildered by the gaps in fossil records of living organisms. **Punctuated equilibrium** is the model of evolution that states that organismal form diverges and species form rapidly over relatively short periods of geological history, and then progress through long stages of stasis with little or no change. Punctuationalists use fossil records to support their claim. It is probable that both gradualism and punctuated equilibrium are correct, depending on the particular lineage studied.

Skill 16.5   Analyzing factors that lead to speciation (e.g., geographic and reproductive isolation, genetic drift)

The most commonly used species concept is the **Biological Species Concept (BSC)**.

This concept states that a species is a reproductive community of populations that occupy a specific niche in nature. It focuses on reproductive isolation of populations as the primary criterion for recognition of species status. The biological species concept does not apply to organisms that are asexual in their reproduction, fossil organisms, or distinctive populations that hybridize.

Reproductive isolation is caused by any factor that impedes two species from producing viable, fertile hybrids. Reproductive barriers can be categorized as **prezygotic** (premating) or **postzygotic** (postmating).

The prezygotic barriers include:

1. **Habitat isolation** – species occupy different habitats in the same territory.

2. **Temporal isolation** – populations reaching sexual maturity/flowering at different times of the year.

3. **Ethological isolation** – behavioral differences that reduce or prevent interbreeding between individuals of different species (including pheromones and other attractants).

4. **Mechanical isolation** – structural differences that make gamete transfer difficult or impossible.

5. **Gametic isolation** – male and female gametes do not attract each other; no fertilization.

The postzygotic barriers include:

1. **Hybrid inviability** – hybrids die before sexual maturity.

2. **Hybrid sterility** – disrupts gamete formation; no normal sex cells.

3. **Hybrid breakdown** – reduces viability or fertility in progeny of the F_2 backcross.

Geographical isolation can also lead to the origin of species. **Allopatric speciation** is speciation without geographic overlap. It is the accumulation of genetic differences through division of a species' range, either through a physical barrier separating the population or through expansion by dispersal. In **sympatric speciation**, new species arise within the range of parent populations. Populations are sympatric if their geographical range overlaps. This usually involves the rapid accumulation of genetic differences (usually chromosomal rearrangements) that prevent interbreeding with adjacent populations.

COMPETENCY 17.0   UNDERSTAND THE PRINCIPLES OF TAXONOMY AND THE RELATIONSHIP BETWEEN TAXONOMY AND THE HISTORY OF EVOLUTION

Skill 17.1   Analyzing criteria used to classify organisms (e.g., morphology, biochemical comparisons)

Scientists estimate that there are more than ten million different species of living things. Of these, 1.5 million have been named and classified. Systems of classification show similarities and assist scientists with a worldwide system of organization.

Carolus Linnaeus is termed the father of taxonomy. **Taxonomy** is the science of classification. Linnaeus based his system on morphology (study of structure). Later on, evolutionary relationships (phylogeny) were also used to sort and group species. The modern classification system uses binomial nomenclature, a two-word name for every species. The genus is the first part of the name and the species is the second part. Notice in the levels explained below that *Homo sapiens* is the scientific name for humans. Starting with the kingdom, the groups get smaller and more alike as one moves down the levels in the classification of humans:

Kingdom: Animalia, Phylum: Chordata, Subphylum: Vertebrata, Class: Mammalia, Order: Primate, Family: Hominidae, Genus: Homo, Species: sapiens

Species are defined by the ability to successfully reproduce with other members of their species. Several different morphological criteria are used to classify organisms:

1. **Ancestral characters** - characteristics that are unchanged after evolution (e.g. 5 digits on the hand of an ape).

2. **Derived characters** - characteristics that have evolved more recently (e.g. the absence of a tail on an ape).

3. **Conservative characters** - traits that change slowly.

4. **Homologous characters** - characteristics with the same genetic basis but used for a different function. (e.g., wing of a bat, arm of a human. The bone structure is the same, but the limbs are used for different purposes).

5. **Analogous characters** – structures that differ, but used for similar purposes (e.g. the wing of a bird and the wing of a butterfly).

6. **Convergent evolution** - development of similar adaptations by organisms that are unrelated.

Biological characteristics are also used to classify organisms. Protein comparison, DNA comparison, and analysis of fossilized DNA are powerful comparative methods used to measure evolutionary relationships between species. Taxonomists consider the organism's life history, biochemical (DNA) makeup, behavior, and geographical distribution. The fossil record is also used to show evolutionary relationships.

Skill 17.2   Interpreting a given phylogenetic tree or cladogram of related species

The typical graphic product of a classification is a **phylogenetic tree**, which represents a hypothesis of the relationships based on branching of lineages through time within a group.

Every time you see a phylogenetic tree, you should be aware that it is making statements on the degree of similarity between organisms, or the particular pattern in which the various lineages diverged (phylogenetic history).

Cladistics is the study of phylogenetic relationships of organisms by analysis of shared, derived character states. Cladograms are constructed to show evolutionary pathways. Character states are polarized in cladistic analysis to be plesiomorphous (ancestral features), symplesiomorphous (shared ancestral features), apomorphous (derived features), and synapomorphous (shared, derived features).

Skill 17.3   Demonstrating the ability to design and use taxonomic keys (e.g., dichotomous keys)

A dichotomous key is a biological tool for identifying unknown organisms to some taxonomic level. It consists of a series of couplets, each consisting of two statements describing characteristics of a particular organism or group of organisms. A choice between the statements is made that bet fits the organism in question. The statements typically begin with broad characteristics and become narrower as more choices are needed.

Example A – Numerical key
1   Seeds round – soybeans
1   Seeds oblong – 2
2   Seeds white – northern beans
2   Seeds black – black beans

Example B - Alphabetical key
A   Seeds oblong – B
A   Seeds round - soybeans
B   Seeds white – northern beans
B   Seeds black – black beans

Skill 17.4   Demonstrating the ability to design and use taxonomic keys (e.g., dichotomous keys)

A taxonomic key, also known as a dichotomous key, allows the identification of objects and the classification of a set of organisms into species-level groups. Taxonomic keys divide objects or organisms progressively by offering two (or more) options for each observed characteristic. The answer to each successive question in the key divides the objects or organisms into two (or more) mutually exclusive groups and determines the next step in the classification process. The process continues until the objects or organisms (or their species/group) are identified.

For example, the following is a basic taxonomic key used to divide and identify five everyday objects, a rubber eraser, a nail, a rubber band, a writing pen, and a penny.

1. Contains metal: go to 2
1. Does not contain metal: go to 3

2. Flat, disc-shaped: penny
2. Not flat or disc-shaped: nail

3. Does not contain rubber: writing pen
3. Contains rubber: go to 4

4. Ring-shaped: rubber band
4. Not ring-shaped: rubber eraser

Note that the key positively identifies the objects by separating them into two mutually exclusive groups at each step.

Scientists design taxonomic keys, similar in structure to the preceding example, to identify the species of observed organisms. For example, field biologists may use a taxonomic key that identifies oak trees based on leaf structure to assign oak trees observed in the field to their appropriate species.

Skill 17.5   Relating changes in the structure and organization of the classification system to developments in biological thought (e.g., evolution, modern genetics)

The current five-kingdom system separates prokaryotes from eukaryotes. The prokaryotes belong to the Kingdom Monera while the eukaryotes belong to Kingdoms Protista, Plantae, Fungi, or Animalia. Recent comparisons of nucleic acids and proteins between different groups of organisms have led to problems concerning the five-kingdom system. Based on these comparisons, alternative kingdom systems have emerged. Six and eight kingdom systems as well as a three-domain system have been proposed as more accurate classification systems. It is important to note that classification systems evolve as more information regarding characteristics and evolutionary histories of organisms arise.

SUBAREA IV. BIOLOGICAL UNITY AND DIVERSITY AND LIFE PROCESSES

COMPETENCY 18.0   UNDERSTAND THE UNITY AND DIVERSITY OF LIFE, INCLUDING COMMON STRUCTURES AND FUNCTIONS

Skill 18.1   Analyzing characteristics of living organisms (e.g., differences between living organisms and nonliving things)

Life has defining properties. Some of the more important processes and properties associated with life include:

- Order
- Reproduction
- Energy utilization
- Growth and development
- Adaptation to the environment

Every living organism exhibits order on some level. Living organisms reproduce such that life comes from life. This is known as biogenesis. Organisms require energy. They utilize energy to carry out life's basic functions. On some occasions, organisms also make energy. The growth and development of living organisms is directed by DNA. In order to grow, develop, mature, and survive, organisms adapt to their specific environment. They do this through the process of homeostasis, which is the ability to maintain a certain status. Organisms also adapt through response to current stimuli. On a long-term scale populations adapt through evolution.

Skill 18.2   Recognizing levels of organization (e.g., cells, tissues, organs)

Life is highly organized. The organization of living systems builds on levels from small to increasingly larger and complex. All aspects, whether it is a cell or an ecosystem, have the same requirements to sustain life. Life is organized from simple to complex in the following way:

Atoms→molecules→organelles→cells→tissues→organs→organ systems→organism

Atoms are the smallest units that characterize a specific chemical element. Atoms combine to form stable molecules. **Molecules** combine and cooperate to form organelles. **Organelles** are the specialized units within a cell that are assigned duties (examples include the nucleus and mitochondria). Organelles work together so that a cell can function. **Cells** are the building blocks of **tissue**. Connected areas of tissue are known as **organs**. Organs working together make **organ systems**. It is the combination of all of these parts that complete the **organism**.

Skill 18.3   Comparing and analyzing the basic life functions carried out by living organisms (e.g., obtaining nutrients, excretion, reproduction)

Members of the five different kingdoms of the classification system of living organisms often differ in their basic life functions. Here we compare and analyze how members of the five kingdoms obtain nutrients, excrete waste, and reproduce.

Bacteria are prokaryotic, single-celled organisms that lack cell nuclei. The different types of bacteria obtain nutrients in a variety of ways. Most bacteria absorb nutrients from the environment through small channels in their cell walls and membranes (chemotrophs), while some perform photosynthesis (phototrophs). Chemoorganotrophs use organic compounds as energy sources while chemolithotrophs can use inorganic chemicals as energy sources. Depending on the type of metabolism and energy source, bacteria release a variety of waste products (e.g. alcohols, acids, carbon dioxide) to the environment through diffusion.

All bacteria reproduce through binary fission (asexual reproduction) producing two identical cells. Bacteria reproduce very rapidly, dividing or doubling every twenty minutes in optimal conditions. Asexual reproduction does not allow for genetic variation, but bacteria achieve genetic variety by absorbing DNA from ruptured cells and conjugating or swapping chromosomal or plasmid DNA with other cells.

Animals are multicellular, eukaryotic organisms. All animals obtain nutrients by eating food (ingestion). Different types of animals derive nutrients from eating plants, other animals, or both. Animal cells perform respiration that converts food molecules, mainly carbohydrates and fats, into energy. The excretory systems of animals, like animals themselves, vary in complexity. Simple invertebrates eliminate waste through a single tube, while complex vertebrates have a specialized system of organs that process and excrete waste.

Most animals, unlike bacteria, exist in two distinct sexes. Members of the female sex give birth or lay eggs. Some less developed animals can reproduce asexually. For example, flatworms can divide in two and some unfertilized insect eggs can develop into viable organisms. Most animals reproduce sexually through various mechanisms. For example, aquatic animals reproduce by external fertilization of eggs, while mammals reproduce by internal fertilization. More developed animals possess specialized reproductive systems and cycles that facilitate reproduction and promote genetic variation.

Plants, like animals, are multi-cellular, eukaryotic organisms. Plants obtain nutrients from the soil through their root systems and convert sunlight into energy through photosynthesis. Many plants store waste products in vacuoles or organs (e.g. leaves, bark) that are discarded. Some plants also excrete waste through their roots.

More than half of all plant species reproduce by producing seeds from which new plants grow. Depending on the type of plant, flowers or cones produce seeds. Other plants reproduce by spores, tubers, bulbs, buds, and grafts. The flowers of flowering plants contain the reproductive organs. Pollination is the joining of male and female gametes that is often facilitated by movement by wind or animals.

Fungi are eukaryotic, mostly multi-cellular organisms. All fungi are heterotrophs, obtaining nutrients from other organisms. More specifically, most fungi obtain nutrients by digesting and absorbing nutrients from dead organisms. Fungi secrete enzymes outside of their body to digest organic material and then absorb the nutrients through their cell walls.

Most fungi can reproduce asexually and sexually. Different types of fungi reproduce asexually by mitosis, budding, sporification, or fragmentation. Sexual reproduction of fungi is different from sexual reproduction of animals. The two mating types of fungi are plus and minus, not male and female. The fusion of hyphae, the specialized reproductive structure in fungi, between plus and minus types produces and scatters diverse spores.

Protists are eukaryotic, single-celled organisms. Most protists are heterotrophic, obtaining nutrients by ingesting small molecules and cells and digesting them in vacuoles. All protists reproduce asexually by either binary or multiple fission. Like bacteria, protists achieve genetic variation by exchange of DNA through conjugation.

Skill 18.4 Recognizing the role of physiological processes (e.g., active transport) that contribute to homeostasis and dynamic equilibrium

The molecular composition of the immediate environment outside of the organism is not the same as it is inside, and the temperature outside may not be optimal for metabolic activity within the organism. **Homeostasis** is the control of these differences between internal and external environments. There are three homeostatic systems to regulate these differences, osmoregulation, excretion, and thermoregulation.

In animals, the kidneys are the site of osmoregulation. **Osmoregulation** deals with maintenance of the appropriate level of water and salts in body fluids for optimum cellular functions. The nephrons maintain osmoregulation by repeatedly filtering fluid waste and by reabsorbing excess water. Excretion is the elimination of metabolic (nitrogenous) waste from the body in the form of urea. The functional unit of excretion is the nephron, which make up the kidneys.

Thermoregulation maintains the internal, or core, body temperature of the organism within a tolerable range for metabolic and cellular processes. Common indications of change in body temperature are shivering and sweating. Both are complex behaviors. The site for thermoregulation is the brain. It is the reception site for many hormones and therefore acts as a processing area. It integrates nerve impulses and commands activity. The heart functions to pump blood through the body, carrying life sustaining nutrients to the cells.

Skill 18.5   Recognizing the relationship of structure and function in all living things

Structure correlates with function in living organisms. More specifically, structure follows function. By analyzing a biological structure you can postulate its function. For example, the bones in a bird's wings have a strong, light honeycomb structure. One can theorize that this structure contributes to the flight of the bird. Another example is the incisors of meat eating animals. These sharp teeth allow for ease in tearing off flesh from their prey.

Finally, at the subcellular level, the extensive folding of the inner membrane of the mitochondria allows for a greater amount of this membrane to fit into the very small organelle.

COMPETENCY 19.0   UNDERSTAND THE GENERAL CHARACTERISTICS, FUNCTIONS, AND ADAPTATIONS OF PRIONS, VIRUSES, BACTERIA, PROTOCTISTS (PROTISTS), AND FUNGI

Skill 19.1   Comparing the structure and processes of prions and viruses to cells

Microbiology includes the study of monera, protists, and viruses. Although **viruses** are not classified as living things, they greatly affect other living things by disrupting cell activity. Viruses are obligate parasites because they rely on the host for their own reproduction. Viruses are composed of a protein coat and a nucleic acid, either DNA or RNA. A bacteriophage is a virus that infects a bacterium. Animal viruses are classified by the type of nucleic acid, presence of RNA replicase, and presence of a protein coat.

There are two types of viral reproductive cycles:

1. **Lytic cycle** - The virus enters the host cell and makes copies of its nucleic acids and protein coats and reassembles. It then lyses or breaks out of the host cell and infects other nearby cells, repeating the process.

2. **Lysogenic cycle** - The virus may remain dormant within the cell until some factor activates it and stimulates it to break out of the cell. Herpes is an example of a lysogenic virus.

Prions are protein fibrils that contain no DNA or RNA. Prions cause scrapie in sheep and bovine spongiform encephalitis ("mad-cow" disease) in cows. This disease is characterized by a sponge-like brain. Prions cause slow developing disease of the nervous system in humans similar to those in animals. Kreutzfeldt-Jacob Syndrome and kuru are two of the human diseases caused by prions. The prion diseases are contracted by consuming the tissue of infected organisms.

Skill 19.2   Comparing archaebacteria and eubacteria

Archaebacteria and eubacteria are the two main branches of prokaryotic (moneran) evolution. Archaebacteria evolved from the earliest cells. Most achaebacteria inhabit extreme environments. There are three main groups of archaebacteria: methanogens, extreme halophiles, and extreme thermophiles. Methanogens are strict anaerobes, extreme halophiles live in high salt concentrations, and extreme thermophiles live in hot temperatures (hot springs).

Most prokaryotes fall into the eubacteria (bacteria) domain. Bacteria are divided according to their morphology (shape). Bacilli are rod shaped bacteria, cocci are round bacteria, and spirilli are spiral shaped bacteria.

The Gram stain is a procedure used to differentiate the cell wall make-up of bacteria. Gram positive bacteria have simple cell walls consisting of large amounts of peptidoglycan. These bacteria pick up the stain, revealing a purple color when observed under the microscope. Gram negative bacteria have a more complex cell wall consisting of less peptidoglycan, but have large amounts of lipopolysaccharides. The lipopolysaccharides resist the stain, revealing a pink color when observed under the microscope. Because of the lipopolysaccharide cell wall, Gram negative bacteria tend to be more toxic and are more resistant to antibiotics and host defense mechanisms.

Bacteria reproduce by binary fission. This asexual process is simply dividing the bacterium in half. All new organisms are exact clones of the parent.

Some bacteria have a sticky capsule that protects the cell wall and is also used for adhesion to surfaces. Pili are surface appendages for adhesion to other cells. Bacteria locomotion is via flagella or taxis. Taxis is the movement towards or away from a stimulus. The methods for obtaining nutrition are, for photosynthetic organisms or producers, the conversion of sunlight to chemical energy, for consumers or heterotrophs, consumption of other living organisms, and, for saprophytes, consumption of dead or decaying material.

In comparison, archaebacteria contain no peptidoglycan in the cell wall, they are not inhibited by antibiotics, they have several kinds of RNA polymerase, and they do not have a nuclear envelope. Eubacteria (bacteria) have peptidoglycan in the cell wall, they are susceptible to antibiotics, they have one kind of RNA polymerase, and they have no nuclear envelope.

Skill 19.3   Analyzing the processes of chromosome and plasmid replication and gene transfer in bacteria

Chromosomal replication in bacteria is similar to eukaryotic DNA replication. A **plasmid** is a small ring of DNA that carries accessory genes separate from those of a bacterial chromosome. Most plasmids in Gram-negative bacteria undergo bidirectional replication, although some replicate unidirectionally because of their small size. Plasmids in Gram-positive bacteria replicate by the rolling circle mechanism.

Some plasmids can transfer themselves (and therefore their genetic information) by a process called conjugation. Conjugation requires cell-to-cell contact. The sex pilus of the donor cell attaches to the recipient cell. Once contact has been established, the transfer of DNA occurs by the rolling circle mechanism.

Skill 19.4   Comparing the structure and function of protoctists (protists)

Protists are the earliest eukaryotic descendants of prokaryotes. Protists are found almost anywhere there is water. Protists can be broadly defined as eukaryotic microorganisms and include the macroscopic algae with only a single tissue type. They are defined by exclusion of characteristics common of the other kingdoms. They are not prokaryotes because they have (usually) a true nucleus and membrane-bound organelles. They are not fungi because fungi lack undulopidia and develop from spores. They are not plants because plants develop from embryos, and they are not animals because animals develop from a blastula.

Most protists have a true (membrane-bound) nucleus, complex organelles (mitochondria, chloroplasts, etc.), an aerobic respiration in mitochondria, and undulipodium (cilia) in some life stage.

The chaotic status of names and concepts of the higher classification of the protists reflects their great diversity of form, function, and life cycles. The protists are often grouped as algae (plant-like), protozoa (animal-like), or fungus-like, based on the similarity of their lifestyle and characteristics to these more clearly defined groups. Two distinctive groups of protists are considered for separation as their own kingdoms. The Archaezoa lack mitochondria, the Golgi apparatus, and have multiple nuclei. The Chromista, including diatoms, brown algae, and "golden" algae with chlorophyll c, have a very different photosynthetic plastid from those found in the green algae and plants.

Skill 19.5   Recognizing the significance of prions, viruses, retroviruses, bacteria, protoctists (protists), and fungi in terms of their beneficial uses or deleterious effects

Although bacteria and fungi may cause disease, they are also beneficial for use as medicines and food. Penicillin is derived from a fungus that is capable of destroying the cell wall of bacteria. Most antibiotics work in this way. Some antibiotics can interfere with bacterial DNA replication or can disrupt the bacterial ribosome without affecting the host cells. Viral diseases have been fought through the use of vaccination, where a small amount of the virus is introduced so the immune system is able to recognize it upon later infection. Antibodies are more quickly manufactured when the host has had prior exposure. Finally, researchers often use virus and retroviruses as vectors for the delivery of genes to bacteria, plants, and animals in genetic engineering processes (e.g., gene therapy, creation of transgenic organisms).

The majority of prokaryotes decompose material for use by the environment and other organisms. The eukaryotic fungi are the most important decomposers in the biosphere. They break down organic material to be used up by other living organisms. The fungi are characterized by a short lived diploid stage, which cannot be viewed except under a microscope. The structures that are visible to the naked eye are typically puffballs, mushrooms, and shelf fungi that represent the dikaryote form of the fungi. The haploid stages are commonly observed as the absorptive hyphae or as asexual reproductive sporangia.

COMPETENCY 20.0   UNDERSTAND THE GENERAL CHARACTERISTICS,
                   LIFE FUNCTIONS, AND ADAPTATIONS OF PLANTS

Skill 20.1   Comparing structures and their functions in nonvascular and
              vascular plants (e.g., mosses, ferns, conifers)

The **non-vascular plants** represent a grade of evolution characterized by
several primitive features for plants: lack of roots, lack of conducting tissues,
reliance on absorption of water that falls on the plant or condenses on the plant
in high humidity, and a lack of leaves.

Non-vascular plants include the liverworts, hornworts, and mosses. Each is
recognized as a separate division.

The characteristics of **vascular plants** are as follows: synthesis of lignin to give
rigidity and strength to cell walls for growing upright, evolution of tracheid cells for
water transport and sieve cells for nutrient transport, and the use of underground
stems (rhizomes) as a structure from which adventitious roots originate.

There are two kinds of vascular plants: non-seeded and seeded. The non-
seeded vascular plant divisions include Division Lycophyta (club mosses),
Division Sphenophyta (horsetails), and Division Pterophyta (ferns). The seeded
vascular plants differ from the non-seeded plants by their method of reproduction,
which we will discuss later. The vascular seed plants are divided into two groups,
the gymnosperms and the angiosperms.

Gymnosperms were the first plants to evolve with the use of seeds for
reproduction, which made them less dependent on water to assist in reproduction.
Their seeds and the pollen from the male are carried by the wind. Gymnosperms
have cones that protect the seeds. Gymnosperm divisions include Division
Cycadophyta (cycads), Division Ginkgophyta (ginkgo), Division Gnetophyta
(gnetophytes), and Division Coniferophyta (conifers).

Angiosperms are the largest group in the plant kingdom. They are the flowering
plants and produce true seeds for reproduction. They arose about seventy
million years ago when the dinosaurs were disappearing. The land was drying
up and the plants' ability to produce seeds that could remain dormant until
conditions became acceptable allowed for their success. They also have more
advanced vascular tissue and larger leaves for increased photosynthesis.
Angiosperms consist of only one division, the Anthrophyta. Angiosperms are
divided into monocots and dicots. Monocots have one cotelydon (seed leaf) and
parallel veins on their leaves. Their flower petals are in multiples of threes.
Dicots have two cotelydons and branching veins on their leaves. Flower petals
are in multiples of fours or fives.

Skill 20.2   Analyzing reproduction and development in the different divisions of plants

Plants accomplish reproduction through alternation of generations. Simply stated, a haploid stage in the plants life history alternates with a diploid stage. The diploid sporophyte divides by meiosis to reduce the chromosome number to the haploid gametophyte generation. The haploid gametophytes undergo mitosis to produce gametes (sperm and eggs). Finally, the haploid gametes fertilize to return to the diploid sporophyte stage.

The non-vascular plants need water to reproduce. The vascular, non-seeded plants reproduce with spores and also need water to reproduce. Gymnosperms use seeds for reproduction and do not require water.

Angiosperms are the most numerous and are therefore the main focus of reproduction in this section.

In a process called **pollination**, plant anthers release pollen grains and animals and the wind carry the grains to plant carpels.

The sperm is released to fertilize the eggs. Angiosperms reproduce through a method of double fertilization. Two sperm fertilize an ovum. One sperm produces the new plant and the other forms the food supply for the developing plant (endosperm). The ovule develops into a seed and the ovary develops into a fruit. Then the wind or animals carry the seeds to new locations and new plants form in a process called **dispersal**.

The development of the egg to form a plant occurs in three stages: growth, morphogenesis (the development of form) and cellular differentiation (the acquisition of a cell's specific structure and function).

Germination

Germination describes the initiation of growth of an organism in a resting stage. In plants, a seedling sprouts or germinates from a seed. Plant seeds contain a stored embryo and food reserves. Upon germination, the plant embryo resumes growth and development and continues to develop into a mature plant. Most types of plant seeds germinate only when environmental conditions, such as light, temperature, and moisture, are optimal. In addition, cellular plant hormones direct the germination and growth processes.

Vegetative propagation

Vegetative propagation refers to the ability of some types of plants to reproduce asexually by producing new plants from existing vegetative structures. Examples of mechanisms of vegetative propagation include the formation of new plants from long underground stems, root sprouting, and budding from leaf edges. Natural vegetative propagation is most common in perennial plants. Vegetative propagation is a form of cloning, because the offspring plants are identical to the parent. In addition, humans often take advantage of the regenerative nature of plants to simplify plant breeding and propagation. Examples of man-made vegetative propagation are cuttings and graftings.

Reproductive Structures

The sporophyte is the dominant phase in reproduction. Sporophytes contain a diploid set of chromosomes and form haploid spores by meiosis. Spores develop into gametophytes that produce gametes by mitosis. Angiosperm reproductive structures are the flowers.

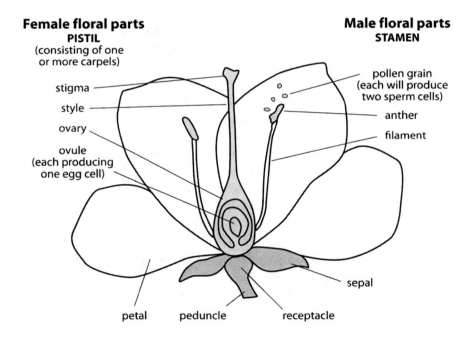

The male gametophytes are pollen grains and the female gametophytes are embryo sacs that are inside of the ovules. The male pollen grains form in the anthers at the tips of the stamens. The ovaries contain the female ovules. Finally, the stamen is the reproductive organ of the male and the carpel is the reproductive organ of the female.

Skill 20.3   Analyzing the structures and forces involved in transport in plants

Roots, stems, leaves, and reproductive structures are the most functionally important parts of plant anatomy. Different types of plants have distinctive anatomical structures. Thus, a discussion of plant anatomy requires an understanding of the classifications of plants.

Roots absorb water and minerals and exchange gases in the soil. Like stems, roots contain xylem and phloem. The xylem transports water and minerals, called xylem sap, upwards. The sugar produced by photosynthesis goes down the phloem in the phloem sap, traveling to the roots and other non-photosynthetic parts of the plant. In addition to water and mineral absorption, roots anchor plants in place preventing erosion by environmental conditions.

Stems are the major support structure of plants. Stems consist primarily of three types of tissue, dermal tissue, ground tissue, and vascular tissue. Dermal tissue covers the outside surface of the stem to prevent excessive water loss and control gas exchange. Ground tissue consists mainly of parenchyma cells and surrounds the vascular tissue providing support and protection. Finally, vascular tissue, xylem and phloem, provides long distance transport of nutrients and water.

Leaves enable plants to capture light and carbon dioxide for photosynthesis. Photosynthesis occurs primarily in the leaves. Plants exchange gases through their leaves via stomata, small openings on the underside of the leaves. Stomata allow oxygen to move in or out of the plant and carbon dioxide to move in. Leaf size and shape varies greatly between species of plants and botanists often identify plants by their characteristic leaf patterns.

Leaf modifications

There are generally 9 types of leaf modifications:

- Scale like - leaf shaped like a scale or awn (Juniper & Cedar)
- Needle - long, slender, tubular or triangular leaf (non-flowering plants like Pine, Fir, etc.)
- Bract like - a modified, often reduced leaf holding a flower or a cluster of flowers, usually colored (Poinsettia)
- Cataphyll - a reduced leaf usually modified for protection, such as bud scales and rhizome scales; its main function is to protect the bud (Pear)
- Storage - leaves are modified for the purpose of storing food (Onion, Garlic, Tulip, Lily)
- Succulent - thick, fleshy leaf, usually modified for water storage found in areas of low rain fall (Crassula, Portulaca, Aloe vera, Sedum)
- Tendril - slender, twining, modified leaf or stem used for clinging to objects for support (Grape, Cucumber)

- Spine - a leaf or leaf part modified into a sharp point (Cactus, Holly)
- Trichome or hair - hair like outgrowths of epidermal cells of leaves (Geranium, Tomato, Chrysanthemum, Peach, Fuzz)

Depending on the necessity, leaves are modified for the survival of plants.

Flowers

Flowers are reproductive organs in plants.

The functions of flowers are two-fold. The brightly colored corolla or petals attract agents of pollination such as insects and birds. The corolla encloses and protects the stamens and pistil

Skill 20.4 Evaluating the evolutionary and adaptive significance of plant structures (e.g., modified leaves, colorful flowers)

Plants require adaptations that allow them to absorb light for photosynthesis. Since they are unable to move about, they must evolve methods to allow them to reproduce successfully. As time passed, the plants moved from a water environment to the land. Advantages of life on land included more available light and a higher concentration of carbon dioxide. Originally, there were no predators and less competition for space on land. Plants had to evolve methods of support, reproduction, respiration, and conservation of water once they moved to land. Reproduction by plants is accomplished through alternation of generations. Simply stated, a haploid stage in the plants life history alternates with a diploid stage.

A division of labor among plant tissues evolved in order to obtain water and minerals from the earth. A wax cuticle is produced to prevent the loss of water. Leaves enabled plants to capture light and carbon dioxide for photosynthesis. Stomata provide openings on the underside of leaves for oxygen to move in or out of the plant and for carbon dioxide to move in. A method of anchorage (roots) evolved. The polymer lignin evolved to give tremendous strength to plants.

COMPETENCY 21.0 UNDERSTAND THE GENERAL CHARACTERISTICS, LIFE FUNCTIONS, AND ADAPTATIONS OF ANIMALS

Skill 21.1   Identifying general characteristics of the embryonic development of invertebrates and vertebrates

Animal tissue becomes specialized during development. The ectoderm (outer layer) becomes the epidermis or skin. The mesoderm (middle layer) becomes muscles and other organs beside the gut. The endoderm (inner layer) becomes the gut, also called the archenteron.

Sponges are the simplest animals and lack true tissue. They exhibit no symmetry.

Diploblastic animals have only two germ layers: the ectoderm and endoderm. They have no true digestive system. Diploblastic animals include the Cnideria (jellyfish). They exhibit radial symmetry.

Triploblastic animals have all three germ layers. Triploblastic animals can be further divided into: Acoelomates, Pseudocoelomates, and Coelomates.

Acoelomates have no defined body cavity. An example is the flatworm (Platyhelminthe), which must absorb food from a host's digestive system.

Pseudocoelomates have a body cavity that is not lined by tissue from the mesoderm. An example is the roundworm (Nematoda).

Coelomates have a true fluid filled body cavity called a coelom derived from the mesoderm. Coelomates can further be divided into protostomes and deuterostomes. In the development of protostomes, the first opening becomes the mouth and the second opening becomes the anus. The mesoderm splits to form the coelom. In the development of deuterostomes, the mouth develops from the second opening and the anus from the first opening. The mesoderm hollows out to become the coelom. Protostomes include animals in the phyla Mollusca, Annelida, and Arthropoda. Deuterostomes include animals in phyla Ehinodermata and Vertebrata.

Development is defined as a change in form. Animals go through several stages of development after fertilization of the egg cell: cleavage, blastula, gastrulation, neuralation, and organogenesis.

Cleavage - the first divisions of the fertilized egg. Cleavage continues until the egg becomes a blastula.

Blastula - a hollow ball of undifferentiated cells.

Gastrulation - the time of tissue differentiation into the separate germ layers, the endoderm, mesoderm, and ectoderm.

Neuralation - development of the nervous system.

Organogenesis - the development of the various organs of the body.

Skill 21.2   Comparing and contrasting the life cycles of invertebrates and vertebrates

The life cycles of invertebrates are relatively simple as opposed to the vertebrates, which are more complex.

The simplest animals like jellyfish reproduce sexually by forming eggs and sperm. These eggs and sperm are released into the water, where fertilization takes place. Some stinging cell animals like hydra reproduce asexually by budding. These buds break off and develop into adults. The worms are one step more complex than the stinging cell animals. Each worm has both male and female reproductive organs and produces eggs and sperm. In sexual reproduction, exchange of eggs and sperm take place with another worm.

In complex invertebrates like insects both sexes are separate and produce eggs and sperm. Females lay the fertilized eggs in large numbers on leaves, where they grow into caterpillars. Insects have a complex life cycle usually consisting of 3 - 4 stages: egg, caterpillar, pupa, and adult. Metamorphosis takes place and the pupa, after resting and undergoing major changes, develops into the adult and the cycle starts again.

In fish, the fertilization of the eggs takes place in water. Fish release millions of eggs and sperm into the water and fertilization occurs in the water. In amphibians, fertilization also takes place in the water. The eggs develop into tadpoles, which live in water, until they develop into adult frogs. Birds and reptiles lay fertilized eggs and take care of them by hatching and guarding them.

Mammals are more complex in their reproductive process. There are egg laying mammals like monotremes, mammals who care for their young in pouches like the kangaroos, and mammals who give birth to fully developed fetuses.

It is very interesting to compare the life cycles of different animals. The simplest are the stinging cell animals and the most complex are the mammals, which give birth to fully developed fetuses.

Skill 21.3   Demonstrating an understanding of physiological processes (e.g., excretion, respiration, aging) of animals and their significance

Animals constantly require oxygen for cellular respiration and need to remove carbon dioxide from their bodies. The respiratory surface must be large and moist. Different animal groups have different types of respiratory organs to perform gas exchange. Some animals use their entire outer skin for respiration (as in the case of worms). Fishes and other aquatic animals have gills for gas exchange. Ventilation increases the flow of water over the gills. This process brings oxygen and removes carbon dioxide through the gills. Fish use a large amount of energy to ventilate its gills. This is because the oxygen available in water is less than that available in the air. The arthropoda (insects) have tracheal tubes that send air to all parts of their bodies.

Gas exchange for smaller insects is provided by diffusion. Larger insects ventilate their bodies by a series of body movements that compress and expand the tracheal tubes. Vertebrates have lungs as their primary respiratory organ. The gas exchange system in all vertebrates is similar to that in humans discussed in a later section.

Osmoregulation and excretion in many invertebrates involves tubular systems. The tubules branch throughout the body. Interstitial fluid enters these tubes and is collected into excretory ducts that empty into the external environment by openings in the body wall. Insects have excretory organs called Malpighian tubes. These organs pump water, salts, and nitrogenous waste into the tubules. These fluids then pass through the hindgut and out the rectum. Vertebrates have kidneys as the primary excretion organ. We describe this system in a later section.

Skill 21.4   Recognizing the relationship between structure and function in given animal species

Porifera - the sponges; they contain spicules for support. Porifera possess flagella for movement in the larval stage, but later become sessile and attach to a firm object. They may reproduce sexually (either by cross or self-fertilization) or asexually (by budding). They are filter feeders and digest food by phagocytosis. They are mostly marine and always live in water due to the need of a hydroskeleton for support.

Cnidaria (Coelenterata) - the jellyfish; these animals possess stinging cells called the nematocyst.

They may be found in a sessile polyp form with the tentacles at the top of the animals or in a moving medusa form with the tentacles floating below. They have a hydroskeleton that uses water for support. They have no true muscles. They may reproduce asexually (by budding) or sexually. They are the first to possess a primitive nervous system.

Platyhelminthes - the flatworms; the flat shape of these animals aid in the diffusion of gases. They are the first group with true muscles. They can reproduce asexually (by regeneration) or sexually. They may be hermaphroditic and possess both sex organs but cannot fertilize themselves. These worms are parasites because they have no true nervous system.

Nematoda - the roundworms; the first animal with a true digestive system with a separate mouth and anus. They may be parasites or simple consumers. They reproduce sexually with male and female worms. They possess longitudinal muscles and thrash about when they move.

Mollusca - clams, octopus; the soft bodied animals. These animals have a muscular foot for movement. They breathe through gills and most are able to make a shell for protection from predators. They have an open circulatory system with sinuses bathing the body regions.

Annelida - the segmented worms; the first with specialized tissue. The circulatory system is more advanced in these worms and is a closed system with blood vessels. The nephridia are their excretory organs. They are hermaphroditic and each worm fertilizes the other upon mating. They support themselves with a hydrostatic skeleton and have circular and longitudinal muscles for movement.

Arthropoda - insects, crustaceans, and spiders; this is the largest group of the animal kingdom. Phylum arthropoda accounts for about 85% of all the animal species. Arthropoda possess an exoskeleton made of chitin. They must molt to grow. They breathe through gills, trachea, or book lungs. Movement varies with members being able to swim, fly, and crawl. There is a division of labor among the appendages (legs, antennae, etc). This is an extremely successful phylum with members occupying diverse habitats.

Echinodermata - sea urchins and starfish; these animals have spiny skin. Their habitat is marine. They have tube feet for locomotion and feeding.

Chordata - all animals with a notocord or a backbone. The classes in this phylum include Agnatha (jawless fish), Chondrichthyes (cartilage fish), Osteichthyes (bony fish), Amphibia (frogs and toads; possess gills that are replaced by lungs during development), Reptilia (snakes, lizards; the first to lay eggs with a protective covering), Aves (birds; warm-blooded), and Mammalia (animals with body hair that bear their young alive, possess mammary glands that produce milk, and are warm-blooded).

Skill 21.5   Analyzing the adaptive and evolutionary significance of animal behaviors and structures

Animal behavior is responsible for courtship leading to mating, communication between species, territoriality, aggression between animals, and dominance within a group. Animal communication is any behavior by one animal that affects the behavior of another animal. Animals use body language, sound, and smell to communicate. Perhaps the most common type of animal communication is the presentation or movement of distinctive body parts. Many species of animals reveal or conceal body parts to communicate with potential mates, predators, and prey. In addition, many species of animals communicate with sound. Examples of vocal communication include the mating "songs" of birds and frogs and warning cries of monkeys. Finally, many animals release scented chemicals called pheromones and secrete distinctive odors from specialized glands to communicate with other animals. Pheromones, chemicals that signal a specific response between members of the same species, are important in reproduction and mating. Glandular secretions of long lasting smells alert animals to the presence of others.

Innate behaviors are inborn or instinctual. An environmental stimulus such as the length of day or temperature results in a behavior. Hibernation among some animals is an innate behavior. **Learned behavior** is modified due to past experience. An example of learned behavior would be an offspring learning to forage by mimicking its parent.

Behavior can be either detrimental or beneficial to an animal. Detrimental behaviors are quickly corrected or the animal perishes. Beneficial behaviors result in increased health for the individual. Health may be defined as the ability to feed, mate, or survive. Healthy individuals translate into healthy populations. This is because it is generally the strongest and most favored individuals who pass on their genes to another generation. Over time, this leads to the evolution of a highly specialized, robust population.

COMPETENCY 22.0   UNDERSTAND THE STRUCTURES AND FUNCTIONS
OF THE HUMAN SKELETAL, MUSCULAR, AND
INTEGUMENTARY SYSTEMS; COMMON
MALFUNCTIONS OF THESE SYSTEMS; AND THEIR
HOMEOSTATIC RELATIONSHIPS WITHIN THE BODY

Skill 22.1   Comparing the structures, locations, and functions of the three types of muscles

The function of the muscular system is to facilitate movement. There are three types of muscle tissue: skeletal, cardiac, and smooth.

Skeletal muscle is voluntary. These muscles are attached to bones and are responsible for their movement. Skeletal muscle consists of long fibers and is striated due to the repeating patterns of the myofilaments (made of the proteins actin and myosin) that make up the fibers.

Cardiac muscle is found in the heart. Cardiac muscle is striated like skeletal muscle, but differs in that plasma membrane of the cardiac muscle causes the muscle to beat even when away from the heart. The action potentials of cardiac and skeletal muscles also differ.

Smooth muscle is involuntary. It is found in organs and enable functions such as digestion and respiration. Unlike skeletal and cardiac muscle, smooth muscle is not striated. Smooth muscle has less myosin and does not generate as much tension as the striated muscles.

Skill 22.2   Demonstrate an understanding of the mechanism of skeletal muscle contraction

The mechanism of skeletal muscle contraction involves a nerve impulse striking a muscle fiber. This causes calcium ions to flood the sarcomere. The myosin fibers creep along the actin, causing the muscle to contract. Once the nerve impulse has passed, calcium is pumped out and the contraction ends.

Skill 22.3   Demonstrate an understanding of the movements of body joints in terms of muscle and bone arrangement and action

The axial skeleton consists of the bones of the skull and vertebrae. The appendicular skeleton consists of the bones of the legs, arms and tail, and shoulder girdle. Bone is a connective tissue.

Parts of the bone include compact bone which gives strength, spongy bone which contains red marrow to make blood cells, yellow marrow in the center of long bones to store fat cells, and the periosteum, which is the protective covering on the outside of the bone.

In addition to bones and muscles, ligaments and tendons are important joint components. A joint is a place where two bones meet. Joints enable movement. Ligaments attach bone to bone. Tendons attach bone to muscle. There are three types of joints:

1. Ball and socket – allow for rotational movement. An example is the joint between the shoulder and the humerus. This joint allows humans to move their arms and legs in many different ways.

2. Hinge – movement is restricted to a single plane. An example is the joint between the humerus and the ulna.

3. Pivot – allows for the rotation of the forearm at the elbow and the hands at the wrist.

Skill 22.4   Relating the structure of the skin to its functions

The skin consists of two distinct layers. The epidermis is the thinner outer layer and the dermis is the thicker, inner layer. Layers of tightly packed epithelial cells make up the epidermis. The tight packaging of the epithelial cells supports the skin's function as a protective barrier against infection.

The top layer of the epidermis consists of dead skin cells and is filled with keratin, a waterproofing protein. The dermis layer consists of connective tissue. It contains blood vessels, hair follicles, sweat glands, and sebaceous glands. An oily secretion called sebum, produced by the sebaceous gland, is released to the outer epidermis through the hair follicles. Sebum maintains the pH of the skin between 3 and 5, which inhibits most microorganism growth.

The skin also plays a role in thermoregulation. Increased body temperature causes skin blood vessels to dilate, resulting in heat radiating from the skin's surface. The sweat glands are also activated, increasing evaporative cooling. Decreased body temperature causes skin blood vessels to constrict. This results in blood from the skin diverting to deeper tissues and reduces heat loss from the surface of the skin.

Skill 22.5	**Demonstrate an understanding of possible causes, effects, prevention, and treatment of malfunctions of the skeletal, muscular, and integumentary systems (e.g., arthritis, skin cancer, scoliosis, osteoporosis)**

Here we consider five common malfunctions of the skeletal, muscular, and integumentary systems: arthritis, skin cancer, scoliosis, osteoporosis, and muscular dystrophy.

Arthritis is disease of the joints that causes pain and loss of movement. Inflammation, pain, and stiffness are the main symptoms of the various types of arthritis. Inflammation is the immune system's response to invasion by foreign bodies or damaged cells and tissue. The symptoms of inflammation (redness, swelling, etc.) result from increased blood flow and fluid leakage into the diseased area caused by chemicals released by immune cells.

Two of the most common types of arthritis are osteoarthritis and rheumatoid arthritis. The cause of osteoarthritis is the gradual breakdown of joint tissue associated with aging and prolonged "wear and tear". The underlying cause of rheumatoid arthritis is unknown. Rheumatoid arthritis is an autoimmune disease in which the body's immune system recognizes healthy tissue (usually joint tissue) as foreign and attacks it.

There is no known way to prevent rheumatoid arthritis, but weight management and avoiding joint injury and over use may prevent or delay the onset of osteoarthritis. The main treatments for arthritis are physical therapy and first and second-line drugs. Regular physical therapy helps maintain joint mobility and range of motion. First-line drugs, such as non-steroidal anti-inflammatory drugs (e.g. aspirin, ibuprofen, and naproxen), corticosteroids, and cox-2 inhibitors (e.g. Celebrex®), provide direct analgesic and anti-inflammatory relief. Second-line drugs used to treat rheumatoid arthritis, such as gold salts, sulfasalazine, methotrexate, chloroquine, hydroxychloroquine, and azathioprine, may delay the progression of the disease symptoms.

Skin cancer is the presence of malignant cells in the outer layers of the skin. The causes of skin cancer are sunburn, UV light damage, and heredity. Skin cancers are changes in the skin, such as growths, non-healing sores, or small lumps. Melanoma, the most dangerous type of skin cancer, can quickly spread to other parts of the body if left untreated. Limiting direct sun exposure and sunburns is the best means of preventing skin cancer. The first-line treatment for skin cancer is excision of the growth or malignancy.

Scoliosis is side-to-side curvature of the spine. The cause of scoliosis is unknown in most cases. Some serious types of scoliosis are caused by spinal birth defects, muscular or nerve damage, and deterioration of bone between the vertebrae. Most minor cases of scoliosis require little more than careful observation while more serious cases may require bracing or surgery. There is no known way to prevent scoliosis, but bracing can prevent progression of the disease.

Osteoporosis is the loss of bone mass leading to brittle bones, neck and back, pain, loss of height, and rounded shoulders. Some causes of osteoporosis are aging, sedentary lifestyle, smoking, calcium and vitamin D deficiency, decreased estrogen levels in women, and long-term use of corticosteroid drugs. Engaging in regular exercise, consuming a diet rich in calcium and vitamin D, refraining from smoking, and limiting caffeine and alcohol consumption may prevent osteoporosis. Hormone replacement treatment is the main treatment for osteoporosis. Other drugs used to treat osteoporosis include biphosphonates, raloxifene, alendronate, and calcitonin.

Muscular dystrophy is a disease characterized by a gradual decrease in muscle size and strength. Muscular dystrophy is caused by wasting of muscle tissue and is a genetic disease. There is no cure for muscular dystrophy and the goal of the main treatments – corticosteroids, physical therapy, and surgery – is management of complications. In addition, there is no known way to prevent muscular dystrophy.

COMPETENCY 23.0   UNDERSTAND THE STRUCTURES AND FUNCTIONS OF THE HUMAN RESPIRATORY AND EXCRETORY SYSTEMS, COMMON MALFUNCTIONS OF THESE SYSTEMS, AND THEIR HOMEOSTATIC RELATIONSHIPS WITHIN THE BODY

Skill 23.1   Demonstrating an understanding of the relationship between surface area and volume and the role of that relationship in the function of the respiratory and excretory systems

The lungs are the respiratory surface of the human respiratory system. A dense net of capillaries contained just beneath the epithelium forms the respiratory surface. The surface area of the epithelium is about $100m^2$ in humans. Based on the surface area, the volume of air inhaled and exhaled is the tidal volume. This is normally about 500mL in adults. Vital capacity is the maximum volume the lungs can inhale and exhale. This is usually around 3400mL.

The kidneys are the primary organs in the excretory system. The kidneys in humans are about 10cm long each. They receive about 20% of the blood pumped with each heartbeat despite their small size. The function of the excretory system is to rid the body of nitrogenous wastes in the form of urea.

Skill 23.2   Analyzing the mechanism of breathing and the process of gas exchange between the lungs and blood and between blood and tissues

The respiratory system functions in the gas exchange of oxygen and carbon dioxide waste. It delivers oxygen to the bloodstream and picks up carbon dioxide for release out of the body. Air enters the mouth and nose, where it is warmed, moistened, and filtered of dust and particles. Cilia in the trachea trap unwanted material in mucus, which can be expelled. The trachea splits into two bronchial tubes and the bronchial tubes divide into smaller and smaller bronchioles in the lungs. The internal surface of the lung is composed of alveoli, which are thin walled air sacs. These allow for a large surface area for gas exchange. The alveoli are lined with capillaries. Oxygen diffuses into the bloodstream and carbon dioxide diffuses out of the capillaries to be exhaled out of the lungs.

The oxygenated blood is carried to the heart and delivered to all parts of the body by hemoglobin, a protein consisting of iron.

The thoracic cavity holds the lungs. The diaphragm muscle below the lungs is an adaptation that makes inhalation possible. As the volume of the thoracic cavity increases, the diaphragm muscle flattens out and inhalation occurs.

Skill 23.3 **Analyzing the role of the kidneys in osmoregulation and waste removal from the blood and the factors that influence nephron function**

The functional unit of excretion is the nephron. There are numerous nephrons that combine to form the kidneys. The structures of the kidney and nephron can be viewed below.

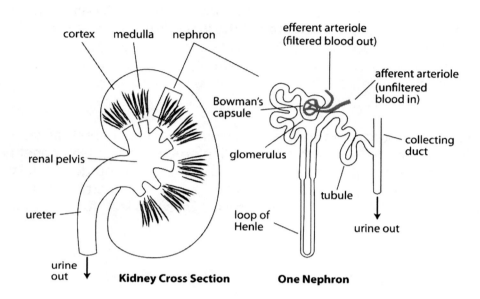

The Bowman's capsule contains the glomerulus, a tightly packed group of capillaries in the nephron. The glomerulus is under high pressure. Due to the high pressure, water, urea, salts and other fluids leak out of the glomerulus and enter the Bowman's capsule. This fluid waste (filtrate) passes through the three regions of the nephron. These are the proximal convoluted tubule, the loop of Henle and the distal tubule. In the proximal convoluted tubule, unwanted molecules are secreted into the filtrate. In the loop of Henle, salt is actively pumped out of the tube and much water is lost due to the hyperosmosity of the inner part (medulla) of the kidney. As the fluid enters the distal tubule, more water is reabsorbed. Urine forms in the collecting duct that leads to the ureter. It then moves to the bladder where it is stored. Urine is passed from the bladder through the urethra. The amount of water reabsorbed back into the body is dependent upon how much fluid an individual has consumed. Urine can be either very dilute or very concentrated depending on the ratio of urea to water.

Skill 23.4 **Demonstrating an understanding of possible causes, effects, prevention, and treatment of malfunctions of the respiratory and excretory systems (e.g., emphysema, nephritis)**

Emphysema is a chronic obstructive pulmonary disease (COPD). These diseases make it difficult for a person to breathe. Airflow through the bronchial tubes is partially blocked making breathing difficult. The primary cause of emphysema is cigarette smoke. People with a deficiency in $alpha_1$-antitrypsin protein production have a greater risk of developing emphysema and at an earlier age. This protein helps protect the lungs from damage done by inflammation. This genetic deficiency is rare and can be tested for in individuals with a family history of the deficiency. There is no cure for emphysema, but there are treatments available. The best prevention against emphysema is to avoid smoking.

Nephritis usually occurs in children. Symptoms include hypertension, decreased renal function, hematuria, and edema. Glomerulonephritis (GN) generally is a more precise term to describe this disease. Nephritis is produced by an antigen-antibody complex that causes inflammation and cell proliferation. Normal kidney tissue is damaged and, if left untreated, nephritis can lead to kidney failure and death.

COMPETENCY 24.0 **UNDERSTAND THE STRUCTURES AND FUNCTIONS OF THE HUMAN CIRCULATORY AND IMMUNE SYSTEMS, COMMON MALFUNCTIONS OF THESE SYSTEMS, AND THEIR HOMEOSTATIC RELATIONSHIPS WITHIN THE BODY**

Skill 24.1 **Demonstrating an understanding of the structure, function, and regulation of the heart and the factors that influence cardiac output**

The function of the closed circulatory system (**cardiovascular system**) is to carry oxygenated blood and nutrients to all cells of the body and return carbon dioxide waste to be expelled from the lungs. The heart, blood vessels, and blood make up the cardiovascular system. The structure of the heart is shown below.

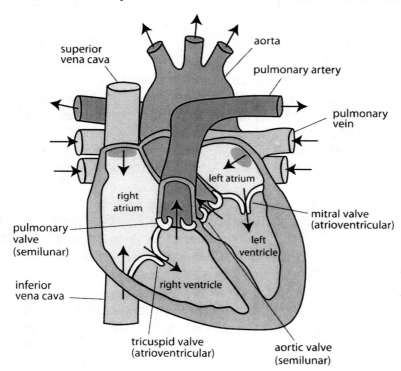

The atria are the chambers that receive blood returning to the heart and the ventricles are the chambers that pump blood out of the heart. There are four valves, two atrioventricular (AV) valves and two semilunar valves. The AV valves are located between each atrium and ventricle. The contraction of the ventricles closes the AV valve to keep blood from flowing back into the atria. The semilunar valves are located where the aorta leaves the left ventricle and the pulmonary artery leaves the right ventricle. The semilunar valves are opened by ventricular contraction to allow blood to be pumped out into the arteries and closed by the relaxation of the ventricles.

The cardiac output is the volume of blood per minute that the left ventricle pumps. This output depends on the heart rate and stroke volume. The **heart rate** is the number of times the heart beats per minute and the **stroke volume** is the amount of blood pumped by the left ventricle each time it contracts. Humans have an average cardiac output of about 5.25 L/min. Heavy exercise can increase cardiac output up to five times. Epinephrine and increased body temperature also increase heart rate and cardiac output.

Cardiac muscle can contract without any signal from the nervous system. It is the sinoatrial node that is the pacemaker of the heart. It is located on the wall of the right atrium and generates electrical impulses that make the cardiac muscle cells contract in unison. The atrioventricular node shortly delays the electrical impulse to ensure the atria empty before the ventricles contract.

Skill 24.2 **Analyzing changes in the circulatory system (e.g., vessel structure and function) and their influence on blood composition and blood flow (e.g., blood cell diversity)**

There are three kinds of blood vessels in the circulatory system: arteries, capillaries, and veins. **Arteries** carry oxygenated blood away from the heart to organs in the body. Arteries branch off to form smaller arterioles in the organs. The arterioles form tiny **capillaries** that reach every tissue. At their downstream end, capillaries combine to form larger venules. Venules combine to form larger **veins** that return blood to the heart. Arteries and veins differ in the direction in which they carry blood.

Blood vessels are lined by endothelium. In veins and arteries, the endothelium is surrounded by a layer of smooth muscle and an outer layer of elastic connective tissue. Capillaries only consist of the thin endothelium layer and its basement membrane that allows for nutrient absorption.

Blood flow velocity decreases as it reaches the capillaries. The capillaries have the smallest diameter of all the blood vessels, but this is not why the velocity decreases. Arteries carry blood to such a large number of capillaries; the blood flow velocity actually decelerates as it enters the capillaries. Blood pressure is the hydrostatic force that blood exerts against the wall of a vessel. Blood pressure is greater in arteries. It is the force that conveys blood from the heart through the arteries and capillaries.

Blood is a connective tissue consisting of the liquid plasma and several kinds of cells. Approximately 60% of the blood is plasma. It contains water salts called electrolytes, nutrients, waste, and proteins. The electrolytes maintain a pH of about 7.4. The proteins contribute to blood viscosity and helps maintain pH. Some of the proteins are immunoglobulins, the antibodies that help fend off infection. Another group of proteins are clotting factors.

The lymphatic system is responsible for returning lost fluid and proteins to the blood. Fluid enters lymph capillaries. This lymph fluid is filtered in the lymph nodes filled with white blood cells that fight off infection.

The two classes of cells in blood are red blood cells and white blood cells. **Red blood cells (erythrocytes)** are the most numerous. They contain hemoglobin, which carries oxygen.

White blood cells (leukocytes) are larger than red blood cells. They are phagocytic and can engulf invaders. White blood cells are not confined to the blood vessels and can enter the interstitial fluid between cells.

There are five types of white blood cells: monocytes, neutrophils, basophils, eosinophils, and lymphocytes.

A third cellular element found in blood is platelets. **Platelets** are made in the bone marrow and assist in blood clotting. The neurotransmitter that initiates blood vessel constriction following an injury is called serotonin. A material called prothrombin is converted to thrombin with the help of thromboplastin. The thrombin is then used to convert fibrinogen to fibrin, which traps red blood cells to form a scab and stop blood flow.

Skill 24.3 Demonstrating an understanding of the possible causes, effects, prevention, and treatment of malfunctions of the circulatory system (e.g., hypertension)

Cardiovascular diseases are the leading cause of death in the United States. Cardiac disease usually results in either a heart attack or a stroke. A heart attack is when cardiac muscle tissue dies, usually from coronary artery blockage. A stroke is when nervous tissue in the brain dies due to the blockage of arteries in the head.

Many heart attacks and strokes are caused by a disease called atherosclerosis. Plaques form on the inner walls of arteries, narrowing the area in which blood can flow. Arteriosclerosis is when the arteries harden from the plaque accumulation. Atherosclerosis can be prevented by a healthy diet that limits lipids and cholesterol and regular exercise. High blood pressure (hypertension) promotes atherosclerosis. Diet, medication, and exercise can reduce high blood pressure and prevent atherosclerosis.

Skill 24.4 **Demonstrating an understanding of the structure, function, and regulation of the immune system (e.g., cell-mediated and humoral responses)**

The immune system is responsible for defending the body against foreign invaders. There are two defense mechanisms: non specific and specific.

The **non-specific** immune mechanism has two lines of defenses. The first lines of defense are the physical barriers of the body. These include the skin and mucous membranes. The skin prevents the penetration of bacteria and viruses as long as there are no abrasions on the skin. Mucous membranes form a protective barrier around the digestive, respiratory, and genitourinary tracts. In addition, the pH of the skin and mucous membranes inhibit the growth of many microbes. Mucous secretions (tears and saliva) wash away many microbes and contain lysozyme that kills many microbes.

The second line of defense includes white blood cells and the inflammatory response. **Phagocytosis** is the ingestion of foreign particles. Neutrophils make up about seventy percent of all white blood cells. Monocytes mature to become macrophages which are the largest phagocytic cells.

Eosinophils are also phagocytic. Natural killer cells destroy the body's own infected cells instead of the invading the microbe directly.

The other second line of defense mechanism is the inflammatory response. The blood supply to the injured area is increased, causing redness and heat. Swelling also typically occurs with inflammation. Histamine is released by basophils and mast cells when the cells are injured. This triggers the inflammatory response.

The **specific** immune mechanism recognizes specific foreign material and responds by destroying the invader. These mechanisms are specific and diverse. They are able to recognize individual pathogens. An **antigen** is any foreign particle that elicits an immune response. An **antibody** is manufactured by the body and recognizes and latches onto antigens, hopefully destroying them. They also have recognition of foreign material versus the self. Memory of the invaders provides immunity upon further exposure.

Immunity is the body's ability to recognize and destroy an antigen before it causes harm. Active immunity develops after recovery from an infectious disease (e.g. chicken pox) or after a vaccination (e.g., mumps, measles, rubella). Passive immunity may be passed from one individual to another and is not permanent. A good example is the immunities passed from mother to nursing child. A baby's immune system is not well developed and the passive immunity they receive through nursing keeps them healthier.

There are two main responses made by the body after exposure to an antigen: humoral and cell-mediated.

1. **Humoral response** - Free antigens activate this response and B cells (lymphocytes from bone marrow) give rise to plasma cells that secrete antibodies and memory cells that will recognize future exposures to the same antigen. The antibodies defend against extracellular pathogens by binding to the antigen and making them an easy target for phagocytes to engulf and destroy. Antibodies are in a class of proteins called immunoglobulins. There are five major classes of immunoglobulins (Ig) involved in the humoral response: IgM, IgG, IgA, IgD, and IgE.

2. **Cell-mediated response** - Cells that have been infected activate T cells (lymphocytes from the thymus). These activated T cells defend against pathogens in the cells or cancer cells by binding to the infected cells and destroying them along with the antigen. T cell receptors on the T helper cells recognize antigens bound to the body's own cells. T helper cells release IL-2 which stimulates other lymphocytes (cytotoxic T cells and B cells). Cytotoxic T cells kill infected host cells by recognizing specific antigens.

Vaccines are antigens given in very small amounts. They stimulate both humoral and cell-mediated responses and help memory cells recognize future exposure to the antigen so antibodies can be produced much faster.

Skill 24.5   Demonstrating an understanding of the possible causes, effects, prevention, and treatment of malfunctions of the immune system (e.g., autoimmune diseases, transplant rejection)

The immune system attacks not only microbes, but also cells that are not native to the host. This is the problem with skin grafts, organ transplantations, and blood transfusions. Antibodies to foreign blood and tissue types already exist in the body. If blood is transfused that is not compatible with the host, these antibodies destroy the new blood cells. There is a similar reaction when tissue and organs are transplanted.

The major histocompatibility complex (MHC) is responsible for the rejection of tissue and organ transplants. This complex is unique to each person. Cytotoxic T cells recognize the MHC on the transplanted tissue or organ as foreign and destroy these tissues. Various drugs are needed to suppress the immune system so this does not happen. The complication with this is that the patient is now more susceptible to infection.

Autoimmune disease occurs when the body's own immune system destroys its own cells. Lupus, Grave's disease, and rheumatoid arthritis are examples of autoimmune diseases. There is no known way to prevent autoimmune diseases. Immunodeficiency is a deficiency in either the humoral or cell mediated immune defenses. HIV is an example of an immunodeficiency disease.

COMPETENCY 25.0   UNDERSTAND HUMAN NUTRITION AND THE STRUCTURES AND FUNCTIONS OF THE HUMAN DIGESTIVE SYSTEM AND ACCESSORY ORGANS, COMMON MALFUNCTIONS OF THE DIGESTIVE SYSTEM, AND ITS HOMEOSTATIC RELATIONSHIPS WITHIN THE BODY

Skill 25.1   Demonstrating an understanding of the roles in the body of the basic nutrients found in foods (e.g., carbohydrates, vitamins, water)

The function of the digestive system is to break food down into nutrients and absorb it into the blood stream where it can be delivered to all cells of the body for use in cellular respiration.

Essential nutrients are those nutrients that the body needs but cannot make. There are four groups of essential nutrients: essential amino acids, essential fatty acids, vitamins, and minerals.

There are ten essential amino acids humans need. A lack of these amino acids results in protein deficiency. There are only a few essential fatty acids.

Vitamins are organic molecules essential for a nutritionally adequate diet. Nutritionists have identified thirteen vitamins essential to humans.

There are two groups of vitamins: water soluble (includes the vitamin B complex and vitamin C) and water insoluble (vitamins A, D and K). Vitamin deficiencies can cause severe problems.

Unlike vitamins, minerals are inorganic molecules. Calcium is needed for bone construction and maintenance. Iron is important in cellular respiration and is a major component of hemoglobin.

Carbohydrates, fats, and proteins are fuel for the generation of ATP. Water is necessary to keep the body hydrated.

Skill 25.2 Demonstrating an understanding of the processes of mechanical and chemical digestion in the digestive system, including contributions of accessory organs

The teeth and saliva begin digestion by breaking food down into smaller pieces and lubricating it so it can be swallowed. The lips, cheeks, and tongue form a bolus or ball of food. It is carried down the pharynx by the process of peristalsis (wave-like contractions) and enters the stomach through the sphincter, which closes to keep food from going back up. In the stomach, pepsinogen and hydrochloric acid form pepsin, the enzyme that hydrolyzes proteins. The food is broken down further by this chemical action and is churned into acid chyme. The pyloric sphincter muscle opens to allow the food to enter the small intestine.

Most nutrient absorption occurs in the small intestine. Its large surface area, accomplished by its length and protrusions called villi and microvilli, allow for a great absorptive surface into the bloodstream. Chyme is neutralized after coming from the acidic stomach to allow the enzymes found there to function. Accessory organs function in the production of necessary enzymes and bile. The pancreas makes many enzymes to break down food in the small intestine. The liver makes bile, which breaks down and emulsifies fatty acids. Any food left after the trip through the small intestine enters the large intestine. The large intestine functions to reabsorb water and produce vitamin K. The feces, or remaining waste, are passed out through the anus.

Skill 25.3 Recognizing the process by which nutrients are transported from inside the small intestine to other parts of the body

The small intestines are where most ingested nutrients are absorbed. Proteins are mainly digested in the stomach, but the small intestines are the site of most chemical digestion, including the break down of peptides into amino acids, the degradation of lipids into fatty acids, and the processing of carbohydrates into simple sugars. Water, electrolytes (sodium, potassium, etc), glucose, amino acids, and fatty acids all enter the body through the small intestines.

The small intestines have extremely high surface area to allow the absorption of large amounts of nutrients. In adult humans, it is approximately 6 meters long and covered in wrinkles called rugae or mucosal folds. The rugae are covered with microscopic projections called villi. Each villi, in turn, is also covered with microvilli, which increases the surface area even further. Together these structures create a surface area of over 250 square meters. On the surface of the villi are enterocytes, a special type of absorptive epithelial cell.

Food is broken down mechanically in the mouth and stomach. Within the small intestines, bile salts from the biliary system and digestive enzymes from the pancreas further digest the food. Contractions in the intestines keep the liberated nutrients in motion and allow them to come into contact with the enterocytes. The various nutrients listed above pass either through tight junctions between the enterocytes (paracellular route) or into the enterocytes themselves through the plasma membranes of the cells (transcellular route).

While some molecules can pass via either route, others must be transported by the transcellular route, taking advantage of special channels in the cellular membranes. The enterocytes maintain a low intracellular concentration of sodium, using ATP-driven sodium pumps. The pumps create both a sodium and a charge gradient across the absorbing surface of the small intestines. The sodium gradient is exploited by a number of co-transporters that move nutrients across the cell membrane. The following briefly summarizes how various nutrients enter.

Water: Passes via either the paracellular or transcellular route, driven entirely by osmosis.

Electrolytes: Typically enter via the transcellular route, often in co-transporters that move both an organic molecule and an ion across the membrane together.

Sugars: Larger sugars are broken down to glucose and other monosaccharides through the various digestive enzymes. A variety of transporters move the sugars across the cell membranes of the enterocytes. Most notably, glucose and galactose are co-transported with sodium ions.

Proteins: Proteins are broken down into very small peptides during digestion (single amino acids, di- and tri-peptides). The transport of these small peptides is analogous to that of sugars. Sodium-dependent transporters exist for acidic, basic, and neutral amino acids and serve as the primary mode of transport.

Fats: Within the small intestines, amphiphatic bile salts surround fat molecules. The hydrophobic regions of the bile salts attach to the triglyceride, while the hydrophilic regions face out. This effectively emulsifies the fat, which then passes directly through the enterocyte membrane or through transport proteins. Once inside the enterocyte, the fat is processed by the endoplasmic reticulum and the Golgi. Triglyceride is packaged with cholesterol, lipoproteins, and other lipids into particles called chylomicrons.

After passing through or between the enterocytes, the nutrients pass into capillaries located within the villi. Fats, however, do not take this path and instead the chylomicrons enter the lymphatic vessels also located in the villi. This drains into the lymphatic system and finally enters the blood. All nutrients then travel through the circulatory system to all the cells in the body.

Skill 25.4 **Demonstrating an understanding of the possible causes, effects, prevention, and treatment of common malfunctions of the digestive system (e.g., ulcers, appendicitis, eating disorders)**

Some of the most common malfunctions of the digestive system include ulcers, appendicitis, and eating disorders. Gastric ulcers are lesions in the stomach lining. Stress used to be blamed for the presence of ulcers, but medical evidence has shown otherwise. Ulcers are caused by bacteria. The bacteria weaken the stomach lining, which is then irritated by stomach acid. Ulcers can be very painful. The most common symptom is severe pain but sufferers may also complain of loss of appetite, weight loss, and bloody bowel movements. Ulcers are often treated with prescription medication to promote their healing.

Appendicitis is the inflammation of the appendix. The appendix has no known function, is open to the intestine, and can be blocked by hardened stool or swollen tissue. The blocked appendix can cause bacterial infections and inflammation leading to appendicitis. The swelling restricts the blood supply, killing the organ tissue. If left untreated, this leads to the rupture of the appendix, allowing the stool and infection to spill out into the abdomen. This condition is life threatening without immediate surgery. Symptoms of appendicitis include lower abdominal pain, nausea, loss of appetite, and fever.

Eating disorders include Anorexia, Bulimia, Binge Eating, and Compulsive Overeating. Anorexia is characterized by excessive weight loss. It involves starvation and is motivated by a desire to be thin. Bulimia is characterized by cycles of binge eating followed by purging. Both anorexia and bulimia can involve strenuous amounts of exercise and/or an inaccurate self-image. These are serious disorders and, if left untreated, can result in death. Dehydration is a common result of Anorexia. Anorexia may also negatively affect a child's growth, bone mass, onset of puberty, and heartbeat. The constant vomiting involved in bulimia can cause severe swelling and damage to of the esophagus. Hospitalization is required if weight loss is too substantial and if intervention on an out-patient basis is not successful. Binge eating and compulsive overeating involve eating too much. The compulsive overeater eats constantly to hide or cope with emotions and issues. The binge eater eats excessively large amounts of food over a small span of time. This is known as a binge and the binges occur periodically. A person suffering from binge eating or compulsive eating is usually overweight. Being overweight puts these individuals at risk for heart and blood pressure conditions. Compulsive overeaters and binge eaters also suffer from low self-esteem and may feel unable to curb their eating cycles.

COMPETENCY 26.0 **UNDERSTAND THE STRUCTURES AND FUNCTIONS OF THE HUMAN NERVOUS AND ENDOCRINE SYSTEMS, COMMON MALFUNCTIONS OF THESE SYSTEMS, AND THEIR HOMEOSTATIC RELATIONSHIPS WITHIN THE BODY**

Skill 26.1 **Demonstrating an understanding of the structures and functions of the central and peripheral nervous systems**

The **central nervous system** (CNS) consists of the brain and spinal cord. The CNS is responsible for the body's response to environmental stimuli. The spinal cord is located inside the spine. It sends out motor commands for movement in response to stimuli. The brain is where responses to more complex stimuli occur. The meninges are the connective tissues that protect the CNS. The CNS contains fluid filled spaces called ventricles. These ventricles are filled when cerebrospinal fluid which is formed in the brain. This fluid cushions the brain and circulates nutrients, white blood cells, and hormones. The CNS's response to stimuli is a reflex. The reflex is an unconscious, automatic response.

The **peripheral nervous system (PNS)** consists of the nerves that connect the CNS to the rest of the body. The sensory division brings information to the CNS from sensory receptors and the motor division sends signals from the CNS to effector cells. The motor division consists of somatic nervous system and the autonomic nervous system. The somatic nervous system is controlled consciously in response to external stimuli. The autonomic nervous system is unconsciously controlled by the hypothalamus of the brain to regulate the internal environment. This system is responsible for the movement of smooth and cardiac muscles as well as the muscles for other organ systems.

The **neuron** is the basic unit of the nervous system. It consists of an axon, which carries impulses away from the cell body to the tip of the neuron; the dendrite, which carries impulses toward the cell body; and the cell body, which contains the nucleus. Synapses are spaces between neurons. Chemicals called neurotransmitters are found close to the synapse. The myelin sheath, composed of Schwann cells, covers the neurons and provides insulation.

Nerve action depends on depolarization and an imbalance of electrical charges across the neuron. A polarized nerve has a positive charge outside the neuron. A depolarized nerve has a negative charge outside the neuron.

Neurotransmitters turn off the sodium pump, which results in depolarization of the membrane. This wave of depolarization (as it moves from neuron to neuron) carries an electrical impulse. This is actually a wave of opening and closing gates that allows for the flow of ions across the synapse. Nerves have an action potential. There is a threshold of the level of chemicals that must be met or exceeded in order for muscles to respond. This is called the "all or none" response.

Skill 26.2   Demonstrating an understanding of the location and function of the major endocrine glands and the function of their associated hormones

The function of the **endocrine system** is to manufacture proteins called hormones. **Hormones** are released into the bloodstream and are carried to a target tissue where they stimulate an action. There are two classes of hormones: steroid and peptide. Steroid hormones come from cholesterol and include the sex hormones. Peptide hormones are derived from amino acids. Hormones are specific and fit receptors on the target tissue cell surface. The receptor activates an enzyme that converts ATP to cyclic AMP. Cyclic AMP (cAMP) is a second messenger from the cell membrane to the nucleus. The genes found in the nucleus turn on or off to cause a specific response.

Hormones are secreted by endocrine cells, which make up endocrine glands. The major endocrine glands and their hormones are as follows:

Hypothalamus – located in the lower brain; signals the pituitary gland.

Pituitary gland – located at the base of the hypothalamus; releases growth hormones and antidiuretic hormone (causing retention of water in kidneys).

Thyroid gland – located on the trachea; lowers blood calcium levels (calcitonin) and maintains metabolic processes (thyroxine).

Gonads – located in the testes of the male and the ovaries of the female; testes release androgens to support sperm formation and ovaries release estrogens to stimulate uterine lining growth and progesterone to promote uterine lining growth.

Pancreas – secretes insulin to lower blood glucose levels and glucagon to raise blood glucose levels.

Skill 26.3   Demonstrating an understanding of the transmission of nerve impulses within and between neurons and the influence of drugs and other chemicals on that transmission

We can divide the drugs that affect our nervous system into three categories: stimulants, depressants, and psychedelic drugs.

1. Stimulants: A stimulant is a drug that speeds up body activities that are controlled by the nervous system. Many stimulants are controlled drugs. Examples are cocaine and amphetamine. Caffeine and nicotine are also stimulants, but they are not controlled drugs.

How does a stimulant speed up body's activities? Neurons carry messages to the brain. Chemical messengers given off by the axon end of one neuron move across the synapse and are picked up by the dendrite end of the next neuron. Usually these chemical messengers are destroyed after crossing the synapse to prevent the original message from going on continuously.

The stimulants may cause the axon of the neuron to give off more of the chemical messenger than normal or the stimulants may prevent the chemical messenger from being destroyed once it reaches the dendrite of the next neuron. In both cases the second neuron keeps receiving chemical messengers. With stimulants, messages move from one neuron to the next for a longer time.

2. Depressants: A depressant slows down messages in the nervous system. They act exactly opposite to stimulants. Their main role is calm behavior. Depressants are controlled drugs. Examples are mild sleep aids, morphine, and barbiturates.

3. Psychedelic drugs: Psychedelics alter the way the mind works and change the signals we receive from our sense organs. Hearing, seeing, and thinking are changed. Senses blend together. Some users may report "tasting colors" and "seeing music". PCP and LSD are examples of psychedelic drugs. Natural psychedelic drugs are those found in certain kinds of mushrooms, cactus plants, marijuana, and the leaves of some desert and jungle plants. Use of PCP causes physical changes including high blood pressure, difficulties with walking and standing, and numbness. Users of PCP may also become violent.

Inhalants are related to psychedelic drugs. Inhalants are drugs that are breathed through the lungs in order to cause a behavior change. The chemicals in glues, paints, and correction fluids are inhalants. Their use can cause irregular heart beat, liver damage and, in some cases, cardiac arrest resulting in death.

Skill 26.4   Evaluating the role of feedback mechanisms in homeostasis (e.g., role of hormones, neurotransmitters)

The thyroid gland produces hormones that help maintain heart rate, blood pressure, muscle tone, digestion, and reproductive functions. The parathyroid glands maintain the calcium level in blood and the pancreas maintains glucose homeostasis by secreting insulin and glucagon. The three gonadal steroids, androgen (testosterone), estrogen, and progesterone, regulate the development of the male and female reproductive organs.

Neurotransmitters are chemical messengers. The most common neurotransmitter is acetylcholine. Acetylcholine controls muscle contraction and heartbeat. A group of neurotransmitters, the catecholamines, include epinephrine and norepinephrine. Epinephrine (adrenaline) and norepinephrine are also hormones. They are produced in response to stress. They have profound effects on the cardiovascular and respiratory systems. These hormones/neurotransmitters can be used to increase the rate and stroke volume of the heart, thus increasing the rate of oxygen delivery to the blood cells.

Skill 26.5   Demonstrating an understanding of the possible causes, effects, prevention, and treatment of malfunctions of the nervous and endocrine systems (e.g., diabetes, brain disorders)

Diabetes is the best known endocrine disorder. Diabetes is caused by a deficiency of insulin resulting in high blood glucose. Type I diabetes is an autoimmune disorder. The immune system attacks the cells of the pancreas, ceasing the ability to produce insulin. Treatment for type I diabetes consists of daily insulin injections. Type II diabetes usually occurs with age and/or obesity. There is usually a reduced response in target cells due to changes in insulin receptors or a deficiency of insulin. Type II diabetics need to monitor their blood glucose levels. Treatment usually includes dietary restrictions and exercise.

Hyperthyroidism is another disorder of the endocrine system, resulting in excessive secretion of thyroid hormones. Symptoms are weight loss, high blood pressure, and high body temperature. The opposite condition, hypothyroidism, causes weight gain, lethargy, and intolerance to cold.

There are many nervous system disorders. Some of the most common include Parkinson's disease, Alzheimer's disease, Epilepsy, Bell's Palsy and tumors and/or cancers of the brain. Parkinson's disease is caused by the degeneration of the basal ganglia in the brain. This causes a breakdown in the transmission of motor impulses to the muscles. Symptoms include tremors, slow movement, and muscle rigidity. Progression of Parkinson's disease occurs in five stages: early, mild, moderate, advanced, and severe. In the severe stage, the person is confined to a bed or chair.

There is no cure for Parkinson's disease. Private research utilizing stem cells is currently underway to find a cure for Parkinson's disease.

Alzheimer's disease is a degenerative disease of the neurological system. Patients suffering from Alzheimer's disease experience impaired memory, thinking, and behavior resulting in confusion, weakened communication skills, and dementia. Alzheimer's disease is progressive, meaning that it worsens with time. Alzheimer's is often associated with frustration and depression, especially in the beginning to mid stages when the patient has lucid moments. There is no cure, but some prescription drugs are available and may help to decrease the symptoms.

Epilepsy is a condition diagnosed by the presence of persistent seizures (note that a single seizure does not indicate epilepsy). The human brain controls and regulates all voluntary and involuntary responses within the body. It does this through nerve cells that normally communicate with each other through electrical activity. A seizure occurs when the brain receives a rush of abnormal electrical signals. These signals temporarily interrupt normal electrical function of the brain. Management of epilepsy strives to control or stop seizure activity. Typical treatments include prescription medications, vagus nervous stimulation (VNS) and surgery.

Bell's Palsy is identified by an otherwise unexplained episode of facial muscle weakness or paralysis. The weakness begins suddenly and worsens over the course of a few days. Bell's Palsy is a result of damage to the facial cranial nerve. Patients experience pain and weakness on just one side of the face or head. This disorder strikes men and women equally and does not distinguish by age. Bell's palsy is not considered permanent and recovery usually begins a month or more after the onset of its symptoms. Treatments include medications for associated symptoms and therapy and protection of the eyes. The majority of people recover fully.

Brain Tumors are abnormal growths of tissue in the brain. The tumor may originate in the brain itself, or originate from another part of the body and travel to the brain. This second type of tumor is said to have metastasized. Brain tumors may be classified as either benign (non-cancerous) or malignant (cancerous). Whether cancerous or non-cancerous, vital functions of the brain can be affected due to the pressure exerted by the growing tumor. The pressure results in the following symptoms: headache, nausea/vomiting, changes in personality, seizure activity, visual difficulty, slurred speech and confusion. Brain tumors are treated by surgery, chemotherapy, radiation, antibiotics and/or rehabilitation depending upon the individual case. Once treated, benign tumors usually do not recur. Brain tumors are fast growing and the prognosis is good if detected early.

COMPETENCY 27.0   **UNDERSTAND THE STRUCTURES AND FUNCTIONS OF THE HUMAN REPRODUCTIVE SYSTEMS, THE PROCESSES OF EMBRYONIC DEVELOPMENT, COMMON MALFUNCTIONS OF THE REPRODUCTIVE SYSTEMS, AND THEIR HOMEOSTATIC RELATIONSHIPS WITHIN THE BODY**

Skill 27.1   **Recognizing the role of hormones in controlling the development and functions of the male and female reproductive systems**

Hormones regulate sexual maturation in humans. Humans cannot reproduce until about the puberty age of 8-14, depending on the individual. The hypothalamus begins secreting hormones that stimulate maturation of the reproductive system and development of the secondary sex characteristics. Reproductive maturity in girls occurs with their first menstruation and occurs in boys with the first ejaculation of viable sperm.

Hormones also regulate reproduction. In males, the primary sex hormones are the androgens, testosterone being the most important. The androgens are produced in the testes and are responsible for the primary and secondary sex characteristics of the male. Female hormone patterns are cyclic and complex. Most women have a reproductive cycle length of about 28 days. The menstrual cycle is specific to the changes in the uterus. The ovarian cycle results in ovulation and occurs in parallel with the menstrual cycle. This parallelism is regulated by hormones. Five hormones participate in this regulation, most notably estrogen and progesterone. Estrogen and progesterone play an important role in the signaling to the uterus and the development and maintenance of the endometruim. Estrogens are also responsible for the secondary sex characteristics of females.

Skill 27.2   **Demonstrating an understanding of gametogenesis, fertilization, and birth control**

Gametogenesis is the production of the sperm and egg cells.

Spermatogenesis begins at puberty in the male. One spermatogonia, the diploid precursor of sperm, produces four sperm. The sperm mature in the seminiferous tubules located in the testes. **Oogenesis**, the production of egg cells (ova), is usually complete by the birth of a female. Egg cells are not released until menstruation begins at puberty. Meiosis forms one ovum with all the cytoplasm and three polar bodies that are reabsorbed by the body. The ovum are stored in the ovaries and released each month from puberty to menopause.

Sperm are stored in the seminiferous tubules in the testes where they mature. Mature sperm are found in the epididymis located on top of the testes. After ejaculation, the sperm travel up the **vas deferens** where they mix with semen made in the prostate and seminal vesicles and travel out the urethra.

Fertilization and embryogenesis

Ovulation releases the egg into the fallopian tubes that are ciliated to move the egg along. Fertilization of the egg by the sperm normally occurs in the fallopian tube. If pregnancy does not occur, the egg passes through the uterus and is expelled through the vagina during menstruation. Levels of progesterone and estrogen stimulate menstruation and are affected by the implantation of a fertilized egg so menstruation will not occur.

Contraception

There are many methods of contraception (birth control) that affect different stages of fertilization. Chemical contraception (birth control pills) prevents ovulation by synthetic estrogen and progesterone. Several barrier methods of contraception are available. Male and female condoms block semen from contacting the egg. Sterilization is another method of birth control. Tubal ligation in women prevents eggs from entering the uterus. A vasectomy in men involves the cutting of the vas deferens. This prevents the sperm from entering the urethra. The most effective method of birth control is abstinence. Worldwide programs have been established to promote abstinence especially amongst teenagers.

Skill 27.3   Demonstrating an understanding of embryonic and fetal development and the potential effects of drugs, alcohol, and nutrition on this process

In this section we will look at some of the potential effects of drugs, alcohol, and nutrition on the process of embryonic and fetal development.

1. Drugs: Cocaine and similar drugs are very detrimental for growth of the fetus. Exposed fetuses often have intrauterine growth retardation, microcephaly, cerebral infarction, urogenital anomalies, an increased risk of sudden infant death syndrome, and neurological and other behavioral abnormalities. These pregnancies are at risk for premature labor, spontaneous abortion, increased perinatal mortality, and fetal death. Drugs are thought to induce birth defects by disrupting the vasculature in the placenta, thereby inducing intrauterine hypoxia and malnutrition.

2. Alcohol: Ethanol is the causative agent of Fetal Alcohol Syndrome (FAS). FAS is seen in approximately 2 in 1000 live births, depending upon culture and socio-economic status. Alcohol is able to permeate the placenta and enter the fetal circulatory system, thereby causing developmental abnormalities. Ethanol impairs placental blood flow to the fetus by constricting blood vessels, inducing hypoxia and fetal malnutrition. Alcohol can cause the following in fetuses: growth deficiencies, micropthalmia, microcephaly, small brain size, and cardiovascular disorders.

3. Nutrition: Nutrition of the mother during pregnancy plays a vital role in the development of the fetus. Malnutrition can lead to a host of complications including brain growth and development.

If fertilization occurs, the zygote begins dividing about 24 hours later. The resulting cells form a blastocyst that implants in about two to three days in the uterus. Implantation promotes secretion of human chorionic gonadotrophin (HCG). This is what is detected in pregnancy tests. The HCG keeps the level of progesterone elevated to maintain the uterine lining in order to feed the developing embryo until the umbilical cord forms.

Organogenesis, the development of the body organs, occurs during the first trimester of fetal development. The heart begins to beat and all the major structures are present at this time. The fetus grows very rapidly during the second trimester of pregnancy. The fetus is about 30 cm long and is very active at this stage. During the third and last trimester, fetal activity may decrease as the fetus grows. Labor is initiated by oxytocin, which causes labor contractions and dilation of the cervix. Prolactin and oxytocin cause the production of milk.

Skill 27.4　Demonstrating an understanding of the possible causes, effects, prevention, and treatment of malfunctions of the reproductive systems (e.g., infertility, birth defects)

The most common diseases or disorders in relation to the reproductive system are infertility, endometriosis, and cancer of the reproductive organs. Infertility is the inability to get pregnant after at least one year of trying. Women who are able to get pregnant but then have repeat miscarriages are also considered infertile.

Causes of infertility in men:

1. Problems making sperm or making very few sperm
2. Problems with the mobility of the sperm in reaching the egg
3. Men born with problems that affect the sperm

The following factors increase the risk of infertility in men: alcohol, drugs, smoking cigarettes, environmental toxins (e.g., pesticides and lead), health problems, medicines, radiation treatment, and chemotherapy).

Causes of infertility in women: ovulation, blocked fallopian tubes due to endometriosis, age, and stress.

To cure infertility, doctors use a variety of treatments:

1. Surgery to correct problems with a woman's ovaries, fallopian tubes, or uterus.

2. Intrauterine insemination (IUI): also known as artificial insemination, in which a woman is inseminated with specially prepared sperm.

3. Assisted reproductive Technology (ART) describes several different methods used to help infertile couples. ART involves removing eggs from a woman, fertilizing them outside the body, and then placing them back into the uterus.

Endometriosis is a disorder causing painful menstruation in a woman. Endometriosis causes endometrial tissue (the tissue lining the uterus) to grow and implant elsewhere in the body, mainly in the abdominal cavity. Some women will take medication to help them cope with symptoms. Occasionally endometriosis affects fertility.

Cancers of the reproductive systems include ovarian cancer, uterine cancer, cervical cancer and prostate cancer. As a society, we have come a long way in our fight against cancer. Survival rates for these cancers are improving. The best way to protect against these cancers is frequent screening. Ovarian cancer has vague symptoms such as abdominal bloating and pelvic discomfort. The most common form of uterine cancer is endometrial cancer, which is a cancer of the inner lining of the uterus. The most common symptom of endometrial cancer is unexpected vaginal bleeding. This cancer has high survival rates when treated early. Risk factors include being over age 50, obesity, high blood pressure, diabetes, and HRT (hormone replacement therapy) without progestin accompaniment. Cervical cancer affects the lining of the cervix, the area joining the uterus and vagina. Pap smears are recommended yearly to check for this slow growing cancer. In addition, a new vaccine has just been approved and is suggested for all 11-12 year old girls.

Prostate cancer affects males only. Prostate cancer has no apparent symptoms in the early stage and only vague symptoms later. According to the Prostate Cancer Foundation, prostate cancer is the most common non-skin cancer in America, affecting 1 in 6 men. The older you are, the more likely you are to be diagnosed with prostate cancer. Many therapies exist, from surveillance only, to radiation, to hormone therapies.

Congenital disorder is a broad and nebulous category. It encompasses the following possibilities:

1. Birth defect: a structural malformation of a body part, recognizable at birth.

2. Congenital physical anomaly: a difference (abnormality) of the structure of a body part, which may or may not be recognized as a problem condition.

3. Congenital malformation: a problem affecting more than one body part.

4. Genetic disorders: may not be apparent at the beginning but will be expressed later in life.

5. Congenital metabolic diseases: also known as inborn errors of metabolism. Most of these are single gene defects.

6. Sporadic birth defects: defects whose causes are not known. They are random and a low recurrence risk for future children.

COMPETENCY 28   UNDERSTAND THE CHARACTERISTICS OF
                   POPULATIONS AND COMMUNITIES AND USE THIS
                   KNOWLEDGE TO INTERPRET POPULATION
                   GROWTH AND INTERACTIONS OF ORGANISMS
                   WITHIN AN ECOSYSTEM

Skill 28.1   Demonstrating an understanding of factors that affect
              population size and growth rate (e.g., carrying capacity,
              limiting factors)

A **population** is a group of individuals of one species that live in the same general
area. Many factors can affect population size and population growth rate. Population
size can depend on the total amount of life a habitat can support. This is the carrying
capacity of the environment. Once the habitat runs out of food, water, shelter, or
space, the carrying capacity decreases, and then stabilizes.

Limiting factors can affect population growth. As a population increases, the
competition for resources is more intense, and the growth rate declines. This is a
density-dependent growth factor. The carrying capacity can be determined by the
density-dependent factor. **Density-independent factors** affect the individuals
regardless of population size. The weather and climate are good examples. Too hot
or too cold temperatures may kill many individuals from a population that has not
reached its carrying capacity.

Skill 28.2   Determining and interpreting population growth curves

Populations grow at different rates depending upon many factors. Zero population
growth rate occurs when the birth and death rates are equal in a population.
Exponential growth rate occurs when there is an abundance of resources and the
growth rate is at its maximum. This is called the intrinsic rate of increase. These
relationships can be explained visually on a growth curve. Two examples can be
viewed on the following page.

An exponentially growing population starts off with little change, then rapidly increases.

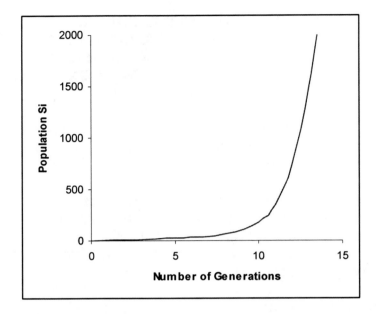

Logistic population growth incorporates the carrying capacity into the growth rate. As a population reaches the carrying capacity, the growth rate begins to slow down and level off.

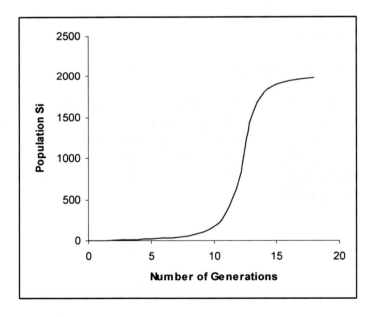

Many populations follow this model of population growth. Humans, however, are an exponentially growing population. Eventually, the carrying capacity of the Earth will be reached, and the growth rate will level off. How and when this will occur remains a mystery.

Skill 28.3 **Analyzing relationships among organisms in a community (e.g., competition, predation, symbiosis)**

There are many interactions that may occur between different species living together. Predation, parasitism, competition, commensalisms, and mutualism are the different types of relationships individuals in a population have with each other.

Predation and **parasitism** result in a benefit for one species and a detriment for the other. Predation is when a predator eats its prey. The common conception of predation is of a carnivore consuming other animals. This is one form of predation. Although not always resulting in the death of the plant, herbivory is a form of predation. Some animals eat enough of a plant to cause death. Parasitism involves a predator that lives on or in its host, causing detrimental effects to the host. Insects and viruses living off and reproducing in their hosts is an example of parasitism. Many plants and animals have defenses against predators. Some plants have poisonous chemicals that will harm the predator if ingested and some animals are camouflaged so they are harder to detect.

Competition is when two or more species in a community use the same resources. Competition is usually detrimental to both populations. Competition is often difficult to find in nature because competition between two populations is not continuous. Either the weaker population will cease to exist, or one population will evolve to utilize other available resources.

Symbiosis is when two species live close together. Parasitism is one example of symbiosis described above. Another example of symbiosis is commensalism. **Commensalism** occurs when one species benefits from the other without harmful effects. **Mutualism** is when both species benefit from the other. Species involved in mutualistic relationships must coevolve to survive. As one species evolves, the other must as well if it is to be successful in life. The grouper fish and a species of shrimp live in a mutualistic relationship. The shrimp feed off parasites living on the grouper. Thus, the shrimp are fed and the grouper stays healthy. Many microorganisms exist in mutualistic relationships.

Skill 28.4   Evaluating the effects of population density on the environment

Population density is the number of individuals per unit area or volume. The spacing pattern of individuals in an area is dispersion. **Dispersion patterns** can be clumped, with individuals grouped in patches; uniform, where individuals are approximately equidistant from each other; or random.

Population densities are usually estimated based on a few representative plots. Aggregation of a population in a relatively small geographic area can have detrimental effects to the environment. Food, water, and other resources will be rapidly consumed, resulting in an unstable environment. A low population density is less harmful to the environment. The use of natural resources will be more widespread, allowing for the environment to recover and continue growth.

COMPETENCY 29.0   UNDERSTAND THE DEVELOPMENT AND STRUCTURE OF ECOSYSTEMS AND THE CHARACTERISTICS OF MAJOR BIOMES

Skill 29.1   Demonstrating an understanding of the flow of energy through the trophic levels of an ecosystem

Trophic levels are based on the feeding relationships that determine energy flow and chemical cycling.

Autotrophs are the primary producers of the ecosystem. **Producers** mainly consist of plants. **Primary consumers** are the next trophic level. The primary consumers are the herbivores that eat plants or algae. **Secondary consumers** are the carnivores that eat the primary consumers. **Tertiary consumers** eat the secondary consumer. These trophic levels may go higher depending on the ecosystem. **Decomposers** are consumers that feed off animal waste and dead organisms. This pathway of food transfer is the food chain.

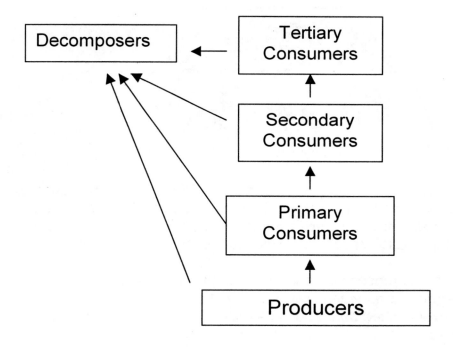

Most food chains are more elaborate, becoming food webs.

Skill 29.2   Comparing the strengths and limitations of various pyramid models (e.g., biomass, numbers, energy)

Energy is lost as the trophic levels progress from producer to tertiary consumer. The amount of energy that is transferred between trophic levels is called the ecological efficiency. The visual of this energy flow is represented in a **pyramid of productivity**.

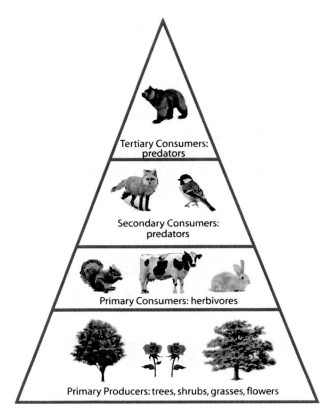

The **biomass pyramid**, depicted above, represents the total dry weight of organisms in each trophic level. A **pyramid of numbers** is a representation of the population size of each trophic level. The producers, being the most populous, are on the bottom of this pyramid with the tertiary consumers on the top with the fewest numbers.

Succession is an orderly process of replacing a community that has been damaged or has begun where no life previously existed. Primary succession occurs where life never existed before, as in a flooded area or a new volcanic island. Secondary succession takes place in communities that were once flourishing but were disturbed by some source, either man or nature, but not totally stripped. A climax community is a community that is established and flourishing.

Abiotic and biotic factors play a role in succession. **Biotic factors** are living things in an ecosystem (e.g., plants, animals, bacteria, and fungi). **Abiotic factors** are non-living aspects of an ecosystem (e.g., soil quality, rainfall, and temperature).

Abiotic factors affect succession by way of the species that colonize the area. Certain species will or will not survive depending on the weather, climate, or soil makeup. Biotic factors such as inhibition of one species due to another may occur. This may be due to some form of competition between the species.

Skill 29.3   Identifying the characteristics and geographic distribution of major biomes

An ecosystem includes all living things in a specific area and the non-living things that affect them. The term **biome** is usually used to classify the general types of ecosystems in the world. Each biome contains distinctive organisms best adapted to that natural environment (including geological make-up, latitude, and altitude). All the living things in a biome are naturally in equilibrium and disturbances in any one element may cause upset throughout the system It should be noted that these biomes may have various names in different areas. For example steppe, savanna, veld, prairie, outback, and scrub are all regional terms that describe the same biome, grassland.

The major terrestrial biomes are desert, grassland, tundra, boreal forest, tropical rainforest, and temperate forest.

Desert

Deserts exist in any location that receives less that 50 cm of precipitation a year. Despite their lack of water and often desolate appearance, the soils, though loose and silty, tend to be rich and specialized plants and animals do populate deserts. Plant species include xerophytes and succulents. Animals tend to be non-mammalian and small (e.g., lizards and snakes). Large animals are not able to find sufficient shade in the desert and mammals, in general, are not well adapted to storing water and withstanding heat. Deserts may be either hot and dry or cold. Hot and dry deserts are what we typically envision when we think of a desert and occur throughout the world. Hot deserts are located in northern Africa, southwestern United States, and the Middle East. Cold deserts have similarly little vegetation and small animals and are located exclusively near the poles in Antarctica, Greenland, much of central Asia, and the Arctic. Both types of deserts do receive precipitation in the winter, though it is in the form of rain in hot deserts and the form of snow in the cold.

Grassland

As the name suggests, grasslands include large expanses of grass with only a few shrubs or trees. There are both tropical and temperate grasslands.

Tropical grasslands cover much of Australia, South America, and India. The weather is warm year-round with moderate rainfall. However, the rainfall is concentrated in half the year and drought and fires are common in the other half of the year. These fires serve to renew rather than destroy areas within tropical grasslands. This type of grassland supports a large variety of animals from insects to mammals both large and small such as squirrels, mice, gophers, giraffes, zebras, kangaroos, lions, and elephants.

Temperate grasslands receive even less rain than tropical grasslands and are found in South Africa, Eastern Europe, and the western United States. As in tropical grasslands, periods of draught and fire serve to renew the ecosystem. Differences in temperature also differentiate the temperate from the tropical grasslands. Temperate grasslands are cooler in general and experience even colder temperatures in winter. These grasslands support similar types of animals as the tropical grasslands: prairie dogs, deer, mice, coyotes, hawks, snakes, and foxes.

The savanna is grassland with scattered individual trees. Plants of the savanna include shrubs and grasses. Temperatures range from 0 - 25 degrees C in the savanna depending on its location. Rainfall is from 90 to 150 cm per year. The savanna is a transitional biome between the rain forest and the desert that is located in central South America, southern Africa, and parts of Australia.

Tundra

Tundras are treeless plains with extremely low temperatures (-28 to 15 degrees C) and little vegetation or precipitation. Rainfall is limited, ranging from 10 to 15 cm per year. A layer of permanently frozen subsoil, called permafrost, is found in the tundra. The permafrost means that no vegetation with deep root systems can exist in the tundra, but low shrubs, mosses, grasses, and lichen are able to survive. These plants grow low and close together to resist the cold temperature and strong winds. The few animals that live in the tundra are adapted to the cold winters (via layers of subcutaneous fat, hibernation, or migration) and raise their young quickly during the summers. Such species include lemmings, caribou, arctic squirrels and foxes, polar bears, cod, salmon, mosquitoes, falcons, and snow birds.

Both arctic and alpine tundra exist, though their characteristics are extremely similar and are distinguished mainly by the location (arctic tundra is located near the North Pole, while alpine tundra is found in the world's highest mountains). Polar tundra or Permafrost - temperature ranges from -40 to 0 degrees C. It rarely gets above freezing. Rainfall is below 10 cm per year. Most water is bound up as ice. Life is limited.

Forests

There are three types of forests, all characterized by the abundant growth of trees, but with difference in climate, flora, and fauna.

Boreal forest (taiga)

These forests are located throughout northern Europe, Asia, and North America, near the poles. The climate typically consists of short, rainy summers followed by long, cold winters with snow. The trees in boreal forests are adapted to the cold winters and are typically evergreens including pine, fir, and spruce. The trees are so thick that there is little undergrowth. A number of animals are adapted to life in the boreal forest, including many mammals such as bear, moose, wolves, chipmunks, weasels, mink, and deer. These coniferous forests have temperatures ranging from -24 to 22 degrees C. Rainfall is from 35 to 40 cm per year. This is the largest terrestrial biome.

Tropical rainforest

Tropical rainforests are located near the equator and are typically warm and wet throughout the entire year. The temperature is constant (25 degrees C) and the length of daylight is about 12 hours. The precipitation is frequent and occurs evenly during the year. In a tropical rainforest, rainfall exceeds 200 cm per year. Tropical rainforests have abundant, diverse species of plants and animals. A tropical dry forest gets scarce rainfall and a tropical deciduous forest has wet and dry seasons. The soil is surprisingly nutrient-poor and most of the biomass is located within the trees themselves. The vegetation is highly diverse including many trees with shallow roots, orchids, vines, ferns, mosses, and palms. Animals are similarly plentiful and varied and include all type of birds, reptiles, bats, insects, and small to medium sized mammals.

Temperate forest

These forests have well defined winters and summers with precipitation throughout the year. Temperate forests are common in Western Europe, eastern North America, and parts of Asia. Common trees include deciduous species such as oak, beech, maple, and hickory. Unlike the boreal forests, the canopy in the temperate forest is not particularly heavy and so various smaller plants occupy the understory. Mammals and birds are the predominate form of animal life. Typical species include squirrels, rabbits, skunks, deer, bobcats, and bear. The temperatures here range from -24 to 38 degrees C. Rainfall is between 65 and 150 cm per year.

Subtypes of temperate forests are Chaparral forests. Chaparral forests experience mild, rainy winters and hot, dry summers. Trees do not grow as well here. Spiny shrubs dominate. Regions include the Mediterranean, the California coastline, and southwestern Australia.

Aquatic ecosystems

Aquatic ecosystems are, as the name suggests, ecosystems located within bodies of water. Aquatic biomes are dived between fresh water and marine. Freshwater ecosystems are closely linked to terrestrial biomes. Lakes, ponds, rivers, streams, and swamplands are examples of freshwater biomes. Marine areas cover 75% of the earth. This biome is organized by the depth of the water. The intertidal zone is from the tide line to the edge of the water. The littoral zone is from the water's edge to the open sea. It includes coral reef habitats and is the most densely populated area of the marine biome. The open sea zone is divided into the epipelagic zone and the pelagic zone. The epipelagic zone receives more sunlight and has a larger number of species.

The ocean floor is called the benthic zone and is populated with bottom feeders. Marine biomes include coral reefs, estuaries, and several systems within the oceans.

Oceans

Within the world's oceans, there are several separate zones, each with it's own temperature profiles and unique species. These zones include intertidal, pelagic, benthic, and abyssal. The interdidal and pelagic zones are further distinguished by the latitude at which they occur (species have evolved to live at in the various temperature water). The intertidal zone is the shore area, which is alternately under and above the water, depending on the tides. Algae, mollusks, snails, crabs, and seaweed are all found in the intertidal zones. The pelagic zone is further from land but near the surface of the ocean. This zone is sometimes called the euphotic zone. Temperatures are much cooler than in the intertidal zone and organisms in this zone include surface seaweeds, plankton, various fish, whales, and dolphins. Further below the ocean's surface is the benthic zone, which is even colder and darker. Much seaweed is found in this zone, as well as, bacteria, fungi, sponges, anemones, sea stars, and some fishes. Deeper still is the abyssal zone, which is the coldest and darkest area of the ocean and has high pressure and low oxygen content. Thermal vents found in the abyssal zone support chemosynthetic bacteria, which are in turn eaten by invertebrates and fishes.

Coral reefs

Coral reefs are located in warm, shallow water near large land masses. The best known example is the Great Barrier Reef off the coast of Australia. The coral itself is the predominant life form in the reefs and obtains its nutrients largely through photosynthesis (performed by the algae). Many other animal life forms also populate coral reefs: many species of fish, octopuses, sea stars, and urchins.

Estuaries

Estuaries are found where fresh and seawater meet, for instance where rivers flow into the oceans. Many species have evolved to thrive in the unique salt concentrations that exist in estuaries. The species include marsh grasses, mangrove trees, oysters, crabs, and certain waterfowl.

Ponds and Lakes

As with the other aquatic biomes, many varied ecosystems occur in ponds and lakes. This is not surprising since lakes vary in size and location. Some lakes are even seasonal, lasting just a few months each year. Additionally, within lakes there are zones, comparable to those in oceans. The littoral zone, located near the shore and at the top of the lake, is the warmest and lightest zone. Organisms in this zone typically include aquatic plants and insects, snails, clams, fish, and amphibians. Further from land, but still at the surface of the lake is the limnetic zone.

Plankton is abundant in the limnetic zone and it is at the bottom of the food chain in this zone, ultimately supporting freshwater fish of all sizes. Deeper in the lake is the profundal zone, which is cooler and darker. Plankton also serves as a valuable food source in this zone since much of it dies and falls to the bottom of the lake. Again, small fish eat this plankton and begin the food chain.

Rivers and Streams

This biome includes moving bodies of water. As expected, the organisms found within streams vary according to latitude and geological features. Additionally, characteristics of the stream change as it flows from its headwaters to the sea. Also, as the depth of rivers increases, zones similar to those seen in the ocean are seen. That is, different species live in the upper, sunlit areas (e.g., algae, top feeding fish, and aquatic insects) and in the darker, bottom areas (e.g., catfish, carp, and microbes).

Wetlands

Wetlands are the only aquatic biome that is partially land-based. They are areas of standing water in which aquatic plants grow. These species, called hydrophytes are adapted to extremely humid and moist conditions and include lilies, cattails, sedges, cypress, and black spruce. Animal life in wetlands includes insects, amphibians, reptiles, many birds, and a few small mammals. Though wetlands are usually classified as a freshwater biome, they are in fact salt marshes that support shrimp, various fish, and grasses.

Skill 29.4   Recognizing the effect of biome degradation and destruction on biosphere stability (e.g., climate changes, deforestation, reduction of species diversity)

Nature replenishes itself continually. Natural disturbances such as landslides and brushfires are not just destructive. Following the destruction, they allow for a new generation of organisms to inhabit the land. For every indigenous organism there exists a natural predator. These predator/prey relationships allow populations to maintain reproductive balance and to not over-utilize food sources, thus keeping food chains in check. Left alone, nature would always find a way to balance itself. Unfortunately, the largest disturbances nature faces are from humans. For example, humans have introduced non-indigenous species to many areas, upsetting the predator/prey relationships. Our building has caused landslides and disrupted waterfront ecosystems. We have damaged the ozone layer and over utilized the land entrusted to us.

The human population has been growing exponentially for centuries. People are living longer and healthier lives than ever before. Better health care and nutrition practices have helped in the survival of the population.

Human activity affects parts of the nutrient cycles by removing nutrients from one part of the biosphere and adding them to another. This results in nutrient depletion in one area and nutrient excess in another. This affects water systems, crops, wildlife, and humans.

Humans are responsible for the depletion of the ozone layer. This depletion is due to chemicals used for refrigeration and aerosols. The consequences of ozone depletion will be severe. Ozone protects the Earth from the majority of UV radiation. An increase of UV will promote skin cancer and unknown effects on wildlife and plants.

Humans have a tremendous impact on the world's natural resources. The world's natural water supplies are affected by human use. Waterways are major sources for recreation and freight transportation. Oil and wastes from boats and cargo ships pollute the aquatic environment. The aquatic plant and animal life is affected by this contamination.

Deforestation for urban development has resulted in the extinction or relocation of several species of plants and animals. Animals are forced to leave their forest homes or perish amongst the destruction. The number of plant and animal species that have become extinct due to deforestation is unknown. Scientists have only identified a fraction of the species on Earth. It is known that if the destruction of natural resources continues, there may be no plants or animals successfully reproducing in the wild.

Humans are continuously searching for new places to form communities. This encroachment on the environment leads to the destruction of wildlife communities. If a biome becomes extinct, the wildlife dies or invades another biome. Preservations established by the government aim at protecting small parts of biomes. While beneficial in the conservation of a few areas, the majority of the environment is still unprotected.

COMPETENCY 30.0 UNDERSTAND THE CONNECTIONS WITHIN AND AMONG THE BIOGEOCHEMICAL CYCLES AND ANALYZE THEIR IMPLICATIONS FOR LIVING THINGS

Skill 30.1 Recognizing the importance of the processes involved in material cycles (e.g., water, carbon, nitrogen, phosphorus)

Biogeochemical cycles are nutrient cycles that involve both biotic and abiotic factors.

Water cycle - Two percent of all the available water is fixed and unavailable in ice or the bodies of organisms. Available water includes surface water (e.g., lakes, oceans, rivers) and ground water (e.g., aquifers, wells). 96% of all available water is from ground water. The water cycle is driven by solar energy. Water is recycled through the processes of evaporation and precipitation. The water present now is the water that has been here since our atmosphere formed.

Carbon cycle - Ten percent of all available carbon in the air (in the form of carbon dioxide gas) is fixed by photosynthesis. Plants fix carbon in the form of glucose. Animals eat the plants and are able to obtain carbon. When animals release carbon dioxide through respiration, the plants again have a source of carbon for further fixation.

Nitrogen cycle - Eighty percent of the atmosphere is in the form of nitrogen gas. Nitrogen must be fixed and taken out of the gaseous form to be incorporated into an organism. Only a few genera of bacteria have the correct enzymes to break the triple bond between nitrogen atoms in a process called nitrogen fixation. These bacteria live within the roots of legumes (e.g., peas, beans, alfalfa) and add nitrogen to the soil so it may be taken up by the plant. Nitrogen is necessary to make amino acids and the nitrogenous bases of DNA.

Phosphorus cycle - Phosphorus exists as a mineral and is not found in the atmosphere. Fungi and plant roots have a structure called mycorrhizae that are able to fix insoluble phosphates into useable phosphorus. Urine and decayed matter return phosphorus to the earth where it can be fixed in the plant. Phosphorus is needed for the backbone of DNA and for ATP manufacturing.

Skill 30.2 Demonstrating an understanding of the role of decomposers in nutrient cycling in ecosystems

Decomposers recycle the carbon accumulated in durable organic material that does not immediately proceed to the carbon cycle. Ammonification is the decomposition of organic nitrogen back to ammonia. This process in the nitrogen cycle is carried out by aerobic and anaerobic bacterial and fungal decomposers. Decomposers add phosphorous back to the soil by decomposing the excretion of animals.

Skill 30.3   Analyzing the role of respiration and photosynthesis in biogeochemical cycling

Biogeochemical cycling is the movement of chemicals between the biotic (living) and abiotic (non-living) parts of an ecosystem. Respiration and photosynthesis play an important role in the cycling of oxygen and carbon. Respiration is the process in which an individual organism uses oxygen and releases carbon dioxide to the atmosphere during energy producing reactions. Photosynthesis, the reverse of respiration, is the process in which an individual plant or microorganism uses the carbon from carbon dioxide to produce carbohydrates with oxygen as a by-product.

Two major forms of carbon in the environment are carbon dioxide gas in the atmosphere and organic macromolecules in living things. Photosynthesis by plants and microorganisms converts carbon dioxide to carbohydrates, removing carbon from the atmosphere and storing it as biomass. Conversely, aerobic and anaerobic respiration by plants, animals, and microorganisms returns carbon to the environment in the form of carbon dioxide or methane gas, respectively.

The main driving force in the oxygen cycle is photosynthesis. Plants and microorganisms perform photosynthesis to produce glucose, releasing oxygen gas to the environment as a by-product. Animals, plants, and microorganisms remove oxygen from the environment, using it to break down glucose in an energy yielding reaction that produces carbon dioxide and water.

Skill 30.4   Evaluating the effects of limiting factors on ecosystem productivity (e.g., light intensity, gas concentrations, mineral availability)

A limiting factor is the component of a biological process that determines how quickly or slowly the process proceeds. Photosynthesis is the main biological process determining the rate of ecosystem productivity, the rate at which an ecosystem creates biomass. Thus, in evaluating the productivity of an ecosystem, potential limiting factors are light intensity, gas concentrations, and mineral availability. The Law of the Minimum states that the required factor in a given process that is most scarce controls the rate of the process.

One potential limiting factor of ecosystem productivity is light intensity because photosynthesis requires light energy. Light intensity can limit productivity in two ways. First, too little light limits the rate of photosynthesis because the required energy is not available. Second, too much light can damage the photosynthetic system of plants and microorganisms thus slowing the rate of photosynthesis. Decreased photosynthesis equals decreased productivity.

Another potential limiting factor of ecosystem productivity is gas concentrations. Photosynthesis requires carbon dioxide. Thus, increased concentration of carbon dioxide often results in increased productivity. While carbon dioxide is often not the ultimate limiting factor of productivity, increased concentration can indirectly increase rates of photosynthesis in several ways. First, increased carbon dioxide concentration often increases the rate of nitrogen fixation (available nitrogen is another limiting factor of productivity). Second, increased carbon dioxide concentration can decrease the pH of rain, improving the water source of photosynthetic organisms.

Finally, mineral availability also limits ecosystem productivity. Plants require adequate amounts of nitrogen and phosphorus to build many cellular structures. The availability of the inorganic minerals phosphorus and nitrogen often is the main limiting factor of plant biomass production. In other words, in a natural environment phosphorus and nitrogen availability most often limits ecosystem productivity, rather than carbon dioxide concentration or light intensity.

COMPETENCY 31.0   UNDERSTAND CONCEPTS OF HUMAN ECOLOGY AND THE IMPACT OF HUMAN DECISIONS AND ACTIVITIES ON THE PHYSICAL AND LIVING ENVIRONMENT

Skill 31.1   Recognizing the importance and implications of various factors (e.g., nutrition, public health, geography, climate) for human population dynamics

The human population has been growing exponentially for centuries. People are living longer and healthier lives than ever before. Better health care and nutrition practices have helped in the survival of the population.

Human activity affects parts of the nutrient cycles by removing nutrients from one part of the biosphere and adding them to another. This results in nutrient depletion in one area and nutrient excess in another. This affects water systems, crops, wildlife, and humans.

Humans are responsible for the depletion of the ozone layer. This depletion is due to chemicals used for refrigeration and aerosols. The consequences of ozone depletion will be severe. Ozone protects the Earth from the majority of UV radiation. An increase of UV will promote skin cancer and unknown effects on wildlife and plants.

Skill 31.2   Predicting the impact of the human use of natural resources (e.g., forests, rivers) on the stability of ecosystems

Humans have a tremendous impact on the world's natural resources. The world's natural water supplies are affected by human use. Waterways are major sources for recreation and freight transportation. Oil and wastes from boats and cargo ships pollute the aquatic environment. The aquatic plant and animal life is affected by this contamination.

Deforestation for urban development has resulted in the extinction or relocation of several species of plants and animals. Animals are forced to leave their forest homes or perish amongst the destruction. The number of plant and animal species that have become extinct due to deforestation is unknown. Scientists have only identified a fraction of the species on Earth. It is known that if the destruction of natural resources continues, there may be no plants or animals successfully reproducing in the wild.

The two categories of natural resources are renewable and nonrenewable. Renewable resources are unlimited because they can be replaced as they are used. Examples of renewable resources are oxygen, wood, fresh water, and biomass. Nonrenewable resources are present in finite amounts or are used faster than they can be replaced in nature. Examples of nonrenewable resources are petroleum, coal, and natural gas.

Strategies for the management of renewable resources focus on balancing the immediate demand for resources with long-term sustainability. In addition, renewable resource management attempts to optimize the quality of the resources. For example, scientists may attempt to manage the amount of timber harvested from a forest, balancing the human need for wood with the future viability of the forest as a source of wood. Scientists attempt to increase timber production by fertilizing, manipulating trees genetically, and managing pests and density. Similar strategies exist for the management and optimization of water sources, air quality, and other plants and animals.

The main concerns in nonrenewable resource management are conservation, allocation, and environmental mitigation. Policy makers, corporations, and governments must determine how to use and distribute scare resources. Decision makers must balance the immediate demand for resources with the need for resources in the future. This determination is often the cause of conflict and disagreement. Finally, scientists attempt to minimize and mitigate the environmental damage caused by resource extraction. Scientists devise methods of harvesting and using resources that do not unnecessarily impact the environment. After the extraction of resources from a location, scientists devise plans and methods to restore the environment to as close to its original state as possible.

Skill 31.3   Recognizing the importance of maintaining biological diversity (e.g., pharmacological products, stability of ecosystems)

Biological diversity is the extraordinary variety of living things and ecological communities interacting with each other throughout the world. Maintaining biological diversity is important for many reasons. First, we derive many consumer products used by humans from living organisms in nature. Second, the stability and habitability of the environment depends on the varied contributions of many different organisms. Finally, the cultural traditions of human populations depend on the diversity of the natural-world.

Many pharmacological products of importance to human health have their origins in nature. For example, scientists first harvested aspirin, a derivative of salicylic acid from the bark of willow trees. In addition, nature is also the source of many medicines including antibiotics, anti-malarial drugs, and cancer fighting compounds. However, scientists have yet to study the potential medicinal properties of many plant species, including the majority of rain forest plants.

Thus, losing such plants to extinction may result in the loss of promising treatments for human diseases.

The basic stability of ecosystems depends on the interaction and contributions of a wide variety of species. For example, all living organisms require nitrogen to live. Only a select few species of microorganisms can convert atmospheric nitrogen into a form that is usable by most other organisms (nitrogen fixation). Thus, humans and all other organisms depend on the existence of the nitrogen-fixing microbes. In addition, the cycling of carbon, oxygen, and water depends on the contributions of many different types of plants, animals, and microorganisms. Finally, the existence and functioning of a diverse range of species creates healthy, stable ecosystems. Stable ecosystems are more adaptable and less susceptible to extreme events like floods and droughts.

Aside from its scientific value, biological diversity greatly affects human culture and cultural diversity. Human life and culture is tied to natural resources. For example, the availability of certain types of fish defines the culture of many coastal human populations. The disappearance of fish populations because of environmental disruptions changes the entire way of life of a group of people. The loss of cultural diversity, like the loss of biological diversity, diminishes the very fabric of the world population.

Skill 31.4 Evaluating methods and technologies that reduce or mitigate environmental degradation

Environmental degradation is damage to an ecosystem, or the biosphere as a whole, resulting from human activities. The underlying causes of environmental degradation are production of energy and consumer products, human population growth and development, and waste disposal. Production of energy and consumer products pollutes the air and contributes to global warming. In addition, harvesting of natural resources can deplete supplies and damage ecosystems. Growth and development of human communities can diminish natural resource supplies, damage the land, and disrupt natural ecosystems. Finally, improper waste disposal can pollute the land and water supplies.

Scientists and policy makers continually attempt to develop and implement new methods and technologies to reduce or mitigate environmental degradation. Cleaner burning fuels or alternative sources of energy that do not pollute the air potential solutions to the energy production-air pollution trade off. In addition, the treatment and filtering of fuel burning by-products can limit environmental impact. However, both developing alternative energy sources and treating current emissions are costly processes. In a market driven economy, governments and policy makers must provide incentives and implement regulations to encourage and require environmental responsibility.

Growth and development of human communities, while inevitable, requires careful planning and attention to environmental concerns. Governmental regulations are often necessary to limit the affect of growth on surrounding ecosystems. Developers and policy makers must attempt to balance the need for increased housing and construction with the importance of respecting and maintaining biodiversity and ecosystem function.

Finally, improper waste disposal can pollute the land and water. Many human and industrial waste products are highly toxic and can cause irreversible environmental damage. Methods of reducing environmental degradation resulting from waste disposal include careful treatment of sewage and human waste, safe disposal of waste products in properly designed locations, and recycling and reuse of waste products.

Skill 31.5   Demonstrating an understanding of the concept of stewardship and ways in which it is applied to the environment

Stewardship is the responsible management of resources. Because human presence and activity has such a drastic impact on the environment, humans are the stewards of the Earth. In other words, it is the responsibility of humans to balance their needs as a population with the needs of the environment and all the Earth's living creatures and resources. Stewardship requires the regulation of human activity to prevent, reduce, and mitigate environmental degradation. An important aspect of stewardship is the preservation of resources and ecosystems for future generations of humans. Finally, the concept of stewardship often, but not necessarily, draws from religious, theological, or spiritual thought and principles.

SUBAREA VII.	**FOUNDATIONS OF SCIENTIFIC INQUIRY:**
	CONSTRUCTED RESPONSE ASSIGNMENT

The content addressed by the constructed-response assignment is described in Subarea I, Objectives 01-06.

Discuss the scientific process.

Best Response:

Science is a body of knowledge that is systematically derived from study, observations, and experimentation. Its goal is to identify and establish principles and theories that may be applied to solve problems. Pseudoscience, on the other hand, is a belief that is not warranted. There is no scientific methodology or application. Some of the more classic examples of pseudoscience include witchcraft, alien encounters, or any topic that is explained by hearsay.

Scientific experimentation must be repeatable. Experimentation leads to theories that can be disproved and are changeable. Science depends on communication, agreement, and disagreement among scientists. It is composed of theories, laws, and hypotheses.

A theory is the formation of principles or relationships that have been verified and accepted.

A law is an explanation of events that occur with uniformity under the same conditions (e.g., laws of nature, law of gravitation).

A hypothesis is an unproved theory or educated guess followed by research to best explain a phenomena. A theory is a proven hypothesis.

Science is limited by the available technology. An example of this would be the relationship of the discovery of the cell to the invention of the microscope. As our technology improves, more hypotheses will become theories and possibly laws. Science is also limited by the data that we can collect. Data may be interpreted differently on different occasions. Science limitations cause explanations to be changeable as new technologies emerge.

The first step in scientific inquiry is posing a question. Next, a hypothesis is formed to provide a plausible explanation. An experiment is then proposed and performed to test this hypothesis. A comparison between the predicted and observed results is the next step. Conclusions are then formed and it is determined whether the hypothesis is correct or incorrect. If incorrect, the next step is to form a new hypothesis and repeat the process.

Better Response:

Science is derived from study, observations, and experimentation. The goal of science is to identify and establish principles and theories that may be applied to solve problems. Scientific theory and experimentation must be repeatable. It is also possible to disprove or change a theory. Science depends on communication, agreement, and disagreement among scientists. It is composed of theories, laws, and hypotheses.

A theory is a principle or relationship that has been verified and accepted through experiments. A law is an explanation of events that occur with uniformity under the same conditions. A hypothesis is an educated guess followed by research. A theory is a proven hypothesis.

Science is limited by the available technology. An example of this would be the relationship of the discovery of the cell to the invention of the microscope. The first step in scientific inquiry is posing a question. Next, a hypothesis is formed to provide a plausible explanation. An experiment is then proposed and performed to test this hypothesis. A comparison between the predicted and observed results is the next step. Conclusions are then formed and it is determined whether the hypothesis is correct or incorrect. If incorrect, the next step is to form a new hypothesis and repeat the process.

Basic Response:

Science is composed of theories, laws, and hypotheses. The first step in scientific inquiry is posing a question. Next, a hypothesis is formed to provide a plausible explanation. An experiment is then proposed and performed to test this hypothesis. A comparison between the predicted and observed results is the next step. Conclusions are then formed and it is determined whether the hypothesis is correct or incorrect. If incorrect, the next step is to form a new hypothesis and repeat the process. Science is always limited by the available technology.

Sample Test

Directions: Read each item and select the best response.

1.   **One common thing found in math, science, and technology is equilibrium. Which of the following is not an example of dynamic equilibrium?**
(Rigorous) (Skill 1.1)

A. a stable population
B. a symbiotic pair of organisms
C. osmoregulation
D. maintaining head position while walking

2.   **Which of the following is not a commonly used cross-discipline concept?**
(Rigorous) (Skill 1.2)

A. algebraic substitution of variables
B. data manipulation
C. graphical interpretation of data
D. diversity

3.   **Which of the following limit the development of technological design and solutions?**

I.   monetary cost
II. time
III. laws of nature
IV. governmental regulation

(Average Rigor) (Skill 1.3)

A. I and II
B. I, II, and IV
C. II and III
D. I, II, and III

4.   **Computer-linked probes automatically gather data and present it in an accessible format. This eliminates the need for:**
(Easy) (Skill 1.4)

A. scientific inquiry
B. constant human observation
C. data analysis
D. experimentation

5.   **The individual parts of cells are best studied using a(n)**
(Average Rigor) (Skill 2.1)

A. ultracentrifuge
B. phase-contrast microscope.
C. CAT scan
D. electron microscope

6.   Which of the following
      made the discovery of
      penicillin?
      (Rigorous) (Skill 2.2)

      A. Pierre Curie
      B. Becquerel
      C. Louis Pasteur
      D. Alexander Fleming

7.   The demand for genetically
      enhanced crops has
      increased in recent years.
      Which of the following is
      not a reason for this
      increased demand?
      (Easy) (Skill 2.3)

      A. Fuel Sources
      B. Increased Growth
      C. Insect Resistance
      D. Better-Looking   Produce

8.   The foremost technique for
      isolating genes is:
      (Rigorous) (Skill 2.4)

      A. restrictive enzyme
         digestion
      B. polymerized chain reaction
      C. northern blot
      D. cDNA cloning

9.   Which is the correct order
      of methodology?

      1. testing a revised
         explanation,
      2. setting up a controlled
         experiment to test an
         explanation,
      3. drawing a conclusion,
      4. suggesting an
         explanation for
         observations,
      5. comparing observed
         results to hypothesized
         results
      (Easy) (Skill 3.1)

      A. 4, 2, 3, 1, 5
      B. 3, 1, 4, 2, 5
      C. 4, 2, 5, 1, 3
      D. 2, 5, 4, 1, 3

10.   The idea that all rational
      beings have intrinsic value
      is the backbone of:
      (Average Rigor) (Skill 3.2)

      A. Utilitarianism
      B. Social Contract Theory
      C. Darwinism
      D. Kantianism

11. As a science fair project, a student cuts a potato into four quarters and plants each in a different type of soil, then administers the same amount of water to each and allows each to receive the same amount of sunlight. Which of the following is the most likely hypothesis being tested? *(Rigorous) (Skill 3.3)*

   A. The different soil environments would effect the growing conditions more than the genetic uniformity of the samples.
   B. The potatoes will grow at the same rate.
   C. One half (two pieces) of the potato will grow faster than the other half.
   D. No growth will occur in the majority of the soils.

12. A student designed a science project testing the effects of light and water on plant growth. You would recommend that she: *(Average Rigor) (Skill 3.4)*

   A. manipulate the temperature as well.
   B. also alter the pH of the water as another variable
   C. omit either water or light as a variable
   D. also alter the light concentration as another variable.

13. A scientific theory: *(Average Rigor) (Skill 4.1)*

   A. proves scientific accuracy
   B. is never rejected.
   C. results in a medical breakthrough.
   D. may be altered at a later time.

14. Which graphical representation would best compare the rate of activity of different enzymes at various temperatures? *(Rigorous) (Skill 4.2)*

   A. Pie Chart
   B. Histogram
   C. Bar Graph
   D. Line Graph

15. In a list of data, the value of the item that occurs the most often is: *(Average Rigor) (Skill 4.4)*

   A. The Mean
   B. The Median
   C. The Mode
   D. The Range

16. In which of the following situations would a linear extrapolation of data be appropriate?
(Rigorous) (Skill 4.5)

A. Computing the death rate of an emerging disease
B. Computing the number of plant species in a forest over time
C. Computing the rate of diffusion with a constant gradient
D. Computing the population at equilibrium

17. The SI measurement for temperature is on the _____ scale.
(Rigorous) (Skill 5.1)

A. Celsius
B. Farenheit
C. Kelvin
D. Rankine

18. The reading of a meniscus in a graduated cylinder is done at the...
(Average Rigor) (Skill 5.2)

A. top of the meniscus
B. middle of the meniscus
C. bottom of the meniscus
D. closest whole number

19. Electrophoresis separates DNA on the basis of...
(Average Rigor) (Skill 6.1)

A. amount of current
B. molecular size
C. positive charge of the molecule
D. solubility of the gel

20. The "Right to Know" law states:
(Average Rigor) (Skill 6.2)

A. the inventory of toxic chemicals checked against the "Substance List" be available
B. that students are to be informed of alternatives to dissection
C. that science teachers are to be informed of student allergies
D. that students are to be informed of infectious microorganisms used in lab

21. Which of the following is not usually found on the MSDS for a laboratory chemical?
(Rigorous) (Skill 6.3)

A. Melting Point
B. Toxicity
C. Storage Instructions
D. Cost

22. **Which of the following is the least ethical choice for a school laboratory activity?** *(Rigorous) (Skill 6.4)*

A. Dissection of a donated cadaver.
B. Dissection of a preserved fetal pig.
C. Measuring the skeletal remains of birds.
D. Pithing a frog to watch the circulatory system.

23. **Who should be notified in the case of a serious chemical spill?**

I. the custodian
II. the fire department
III. the chemistry teacher
IV. the administration
(Easy) (Skill 6.5)

A. I
B. II
C. II and III
D. II and IV

24. **Which kingdom is comprised of organisms made of one cell with no nuclear membrane?** *(Easy) (Skill 7.1)*

A. Monera
B. Protista
C. Fungi
D. Algae

25. **Which of the following is not considered evidence of Endosymbiotic Theory?** *(Rigorous) (Skill 7.2)*

A. The presence of genetic material in mitochondria and plastids.
B. The presence of rirobosomes within mitochondria and plastids.
C. The presence of a double layered membrane in mitochondria and plastids.
D. The ability of mitochondria and plastids to reproduce.

26. **What is necessary for diffusion to occur?** *(Average Rigor) (Skill 7.3)*

A. carrier proteins
B. energy
C. a concentration gradient
D. a membrane

27. **Negatively charged particles that circle the nucleus of an atom are called...** *(Easy) (Skill 8.1)*

A. neutrons
B. neutrinos
C. electrons
D. protons

28. Which protein structure consists of the coils and folds of polypeptide chains?
(Average Rigor) (Skill 8.2)

A. secondary structure
B. quaternary structure
C. tertiary structure
D. primary structure

29. Which of the following are properties of water?
I. High specific heat
II. Strong hydrogenbonds
III. Good solvent
IV. High freezing point
(Rigorous) (Skill 8.3)

A. I, III, IV
B. II and III
C. I and II
D. II, III, IV

30. Sulfur oxides and nitrogen oxides in the environment react with water to cause:

(Easy) (Skill 8.4)

A. ammonia
B. acidic precipitation
C. sulfuric acid
D. global warming

31. ATP is known to bind to phosphofructokinase-1 (an enzyme that is part of glycolysis). This results in a change in the shape of the enzyme and causes the rate of ATP production to fall. Which answer best explores what has happened?
(Rigorous) (Skill 8.5)

A. binding of a coenzyme
B. an allosteric change in the enzyme
C. competitive inhibition
D. uncompetive inhibition

32. During the Krebs cycle, 8 carrier molecules are formed. What are they?
(Rigorous) (Skill 9.2)

A. 3 NADH, 3 $FADH_2$, 2 ATP
B. 6 NADH and 2 ATP
C. 4 $FADH_2$ and 4 ATP
D. 6 NADH and 2 $FADH_2$

33. The most ATP is generated through...
(Rigorous) (Skill 9.3)

A. fermentation
B. glycolysis
C. chemiosmosis
D. the Krebs cycle

34. **Which of the following is not a factor that effects the rate of both photosynthesis and respiration in plants?** *(Average Rigor) (Skill 9.4)*

   A. the concentration of NADP and FAD
   B. the tempature
   C. the structure of the plants
   D. the availability of the different substrates

35. **The product of anaerobic respiration in animals is:** *(Average Rigor) (Skill 9.5)*

   A. carbon dioxide
   B. lactic acid
   C. pyruvate
   D. ethyl alcohol

36. **Which of the following is not true of both chloroplasts and mitochondria?** *(Easy) (Skill 9.6)*

   A. use of the inner membrane for most of it's activity
   B. converts energy from one form to another
   C. uses an electron transport chain
   D. is an important part of the carbon cycle

37. **DNA synthesis results in a strand that is synthesized continuously. This is the...** *(Average Rigor) (Skill 10.1)*

   A. lagging strand
   B. leading strand
   C. template strand
   D. complementary strand

38. **Which of the following is not posttranscriptional processing?** *(Rigorous) (Skill 10.2)*

   A. 5' capping
   B. intron splicing
   C. polypeptide splicing
   D. 3' polyadenylation

39. **Any change that affects the sequence of bases in a gene is called a(n):** *(Easy) (Skill 10.3)*

   A. deletion
   B. polyploid
   C. mutation
   D. duplication

40. **The lac operon**

   I. **contains the lac Z, lac Y, and lac A genes**
   II. **converts glucose to lactose**
   III. **contains a repressor**
   IV. **is on when the repressor is activated**
   (Average Rigor) (Skill 10.4)

   A. I
   B. II
   C. III and IV
   D. I and III

41. **The polymerase chain reaction**
(Average Rigor) (Skill 11.2)

   A. is a group of polymerases.
   B. is a technique for amplifying DNA.
   C. is a primer for DNA synthesis.
   D. is synthesis of polymerase.

42. **Which is not a way that genetic engineering advancement has been used to create a useful application.**
(Rigorous) (Skill 11.3)

   A. The creation of safer viral vacines.
   B. The creation of bacteria that produce hormones for medical use.
   C. The creation of bacteria to breakdown toxic waste.
   D. The creation of organisms that are successfully being used as sources for alternative fuels.

43. **Which is not a way that cDNA cloning has been used?**
(Rigorous) (Skill 11.4)

   A. to provide evidence for taxonomic organization
   B. to study the mutations that lead to diseases such as hemophilia
   C. to determine the structure of a protein
   D. to understand the methods of gene regulation

44. **Which of the following is not a way in which the U.S. government has regulated genetic engineering?**
(Average Rigor) (Skill 11.6)

   A. stem cell research
   B. whether engineered food may go to market
   C. the isolation and study of genes from nonpathogenic organisms
   D. the use of gene therapy treatments

45. **Identify this stage of mitosis.**

(Average Rigor) (Skill 12.1)

A. anaphase
B. metaphase
C. prophase
D. telophase

46. **Which is not a possible effect of polyploidy?**
(Rigorous) (Skill 12.2)

A. more robust members of a species
B. the creation of a cross species offspring
C. the creation of a new species
D. cells that produce higher levels of desired proteins

47. **Which process(es) result(s) in a haploid chromosome number?**
(Easy) (Skill 12.3)

A. both meiosis and mitosis
B. mitosis
C. meiosis
D. replication and division

48. **Cancer cells divide extensively and invade other tissues. This continuous cell division is due to:**
(Average Rigor) (Skill 12.5)

A. density dependent inhibition
B. density independent inhibition
C. chromosome replication
D. inhibition hormones

49. **The ratio of brown-eyed to blue-eyed children from the mating of a blue-eyed male to a heterozygous brown-eyed female would be expected to be which of the following?**
(Average Rigor) (Skill 13.1)

A. 2:1
B. 1:1
C. 1:0
D. 1:2

50. **Amniocentesis is:**
(Average Rigor) (Skill 13.2)

A. a non-invasive technique for detecting genetic disorders
B. a bacterial infection
C. extraction of amniotic fluid
D. removal of fetal tissue

51. Sutton observed that genes and chromosomes behaved the same. This led him to his theory which stated
(Easy) (Skill 13.3)

A. that meiosis causes chromosome separation.
B. that linked genes are able to separate.
C. that genes and chromosomes have the same function
D. that genes are found on chromosomes

52.

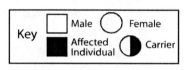

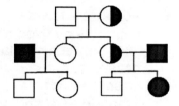

Based on the pedigree chart above, what term best describes the nature of the trait being mapped?
(Rigorous) (Skill 13.4)

A. autosomal recessive
B. sex-linked
C. incomplete dominance
D. co-dominance

53. A women has Pearson Syndrome, a diseasecaused by a mutation in the mitochondrial DNA. Which of the following individuals would you expect to see the disease in?

I   Her Daughter
II Her Son
III Her Daughter's son
IV Her Son's daughter
(Rigorous) (Skill 13.5)

A. I, III
B. I, II, III
C. II, IV
D. I,II, III, IV

54. Which of the following factors will affect the Hardy-Weinberg law of equilibrium, leading to evolutionary change?
(Average Rigor) (Skill 14.1)

A. no mutations
B. non-random mating
C. no immigration or emigration
D. large population

55. **Which of the following is an example of a phenotype that gives the organism an advantage in their home environment?**
(Rigorous) (Skill 14.2)

   A. The color of Pepper Moths in England
   B. Thornless roses in a nature perserve
   C. Albinoism in naked mole rats
   D. Mediterranean fruit fly having a large thorax

56. **All of the following gasses made up the primitive atmosphere except...**
(Average Rigor) (Skill 15.1)

   A. ammonia
   B. methane
   C. oxygen
   D. hydrogen

57. **The first cells that evolved on Earth were probably of which type?**
(Easy) (Skill 15.2)

   A. autotrophs
   B. eukaryotes
   C. heterotrophs
   D. prokaryotes

58. **A fossil of a dinosaur genus known as Saurolophus have been found in both Western North America and in Mongolia. What is the most likely explanation for these findings?**
(Rigorous) (Skill 15.3)

   A. Convergent evolution on the two continents lead to two species of dinosaurs with significant enough similarities to be placed in the same genus.
   B. With the gaps in the fossil record there are currently mutliple competeing theories to explain the presence of these fossils on two separate continents.
   C. A distant ancestor of these dinosaurs evolved at a point before these land masses became separated.
   D. Although Asia and North America were separate continents at the time, low sea levels made it possible for the dinosaurs to walk from one continent to the other.

59. **Which aspect of science does not support evolution?**
(Average Rigor) (Skill 15.4)

A. comparative anatomy
B. organic chemistry
C. comparison of DNA among organisms
D. analogous structures

60. **Which process contributes to the large variety of living things in the world today?**
(Easy) (Skill 16.1)

A. meiosis
B. asexual reproduction
C. mitosis
D. alternation of generations

61. **An animal choosing its mate because of attractive plumage or a strong mating call is an example of:**
(Average Rigor) (Skill 16.2)

A. Mechanical isolation.
B. Natural Selection.
C. Directional Selection.
D. Sexual Selection.

62. **Which of the following best exemplifies the Theory of Inheritance of Acquired Traits?**
(Rigorous) (Skill 16.3)

A. Giraffes need to reach higher for leaves to eat, so their necks stretch. The giraffe babies are then born with longer necks. Eventually, there are more long-necked giraffes in the population.
B. Giraffes with longer necks are able to reach more leaves, so they eat more and have more babies than other giraffes. Eventually, there are more long-necked giraffes in the population.
C. Giraffes want to reach higher for leaves to eat, so they release enzymes into their bloodstream, which in turn causes fetal development of longer-necked giraffes. Eventually, there are more long-necked giraffes in the population.
D. Giraffes with long necks are more attractive to other giraffes, so they get the best mating partners and have more babies. Eventually, there are more long-necked giraffes in the population.

63. The fossil record is often pointed to as evidence of punctuated equilibrium. Which of the following examples can only be explained by punctuated equilibrium and not gradualism?

I Coelacanth fish (once thought extinct) have been relatively unchanged for millions of years.
II The sudden appearance of a large number of different soft-bodied animals around 530 million years ago.
III 10 million year old fossils and modern ginko plants are nearly identical.
IV Fossils of Red Deer for the Island of Jersey show a six-fold decrease in size in body weight in the last 6000 years.
(Rigorous) (Skill 16.4)

A. I, III
B. II, IV
C. I, II, III, IV
D. None of the above

64. The biological species concept applies to
(Average Rigor) (Skill 16.5)

A. asexual organisms
B. extinct organisms
C. sexual organisms
D. fossil organisms

65. Members of the same species:
(Easy) (Skill 17.1)

A. look identical
B. never change
C. reproduce successfully within their group
D. live in the same geographic location

66. What are a majority of phyogenetic trees now being create based upon?
(Average Rigor) (Skill 17.2)

A. analogous structures
B. homologous structures
C. genetic analysis
D. behavioral trends

67.

Using the folowing taxonomic key identify the tree that this branch came from?

1   Are the leaves PALMATELY COMPOUND (BLADES arranged like fingers on a hand)? - 2
1   Are the leaves PINNATELY COMPOUND (BLADES arranged like the veins of a feather)? - 3
2   Are there usually 7 BLADES - *Aesculus hippocastanum*
2   Are there usually 5 BLADES - *Aesculus glabra*
3   Are there mostly 3-5 BLADES that are LOBED or coarsely toothed? - *Acer negundo*
3   Are there mostly 5-13 BLADES with smooth or toothed edges? - *Fraxinus Americana*
(Rigorous) (Skill 17.3)

A. Aesculus hippocastanum
B. Aesculus glabra
C. Acer negundo
D. Fraxinus Americana

68.   Which of the following systems uses Archea (or Archeabacteria) as the most inclusive level of the taxonomic system.

I   Three Domain System
II   Five Kingdom System
III Six Kingdom System
IV   Eight Kingdom System
(Rigorous) (Skill 17.4)

A. II, III
B. I, IV
C. I, III, IV
D. I, II, III, IV

69.   Identify the correct sequence of organization of living things.
(Easy) (Skill 18.2)

A. cell – organelle – organ system – tissue – organ – organism
B. cell – tissue – organ – organ system – organelle – organism
C. organelle – cell – tissue – organ – organ system – organism
D. tissue – organelle – organ – cell – organism – organ system

70. **Which of the following is not a necessary characteristic of living things?**
(Average Rigor) (Skill 18.3)

A. Movement.
B. Reduction of local entropy.
C. Ability to cause change in local energy form.
D. Reproduction.

71. **Which of the following hormones is most involved in the process of osmoregulation?**
(Rigorous) (Skill 18.4)

A. Antidiuretic Hormone.
B. Melatonin.
C. Calcitonin.
D. Gulcagon.

72. **The shape of a cell depends on its...**
(Average Rigor) (Skill 18.5)

A. function
B. structure
C. age
D. size

73. **A virus that can remain dormant until a certain environmental condition causes it's rapid increase is said to be:**
(Average Rigor) (Skill 19.1)

A. lytic
B. benign
C. saprophytic
D. lysogenic

74. **Using a gram staining technique, it is observed that E. coli stains pink. It is therefore...**
(Average Rigor) (Skill 19.2)

A. gram positive
B. dead
C. gram negative
D. gram neutral

75. **Segments of DNA can be transferred from the DNA of one organism to another through the use of which of the following?**
(Rigorous) (Skill 19.3)

A. bacterial plasmids
B. viruses
C. chromosomal sharing
D. single strand DNA

76. Protists are classified into major groups according to
(Average Rigor) (Skill 19.4)

A. their method of obtaining nutrition.
B. reproduction.
C. metabolism.
D. their form and function.

77. Laboratory researchers have classified fungi as distinct from plants because the cell walls of fungi _____
(Rigorous) (Skill 19.5)

A. contain chitin.
B. contain lignin.
C. contain lipopolysaccharides.
D. contain cellulose.

78.

Indentify the correct characteristiscs for the plant pictured above.
(Rigorous) (Skill 20.1)

A. seeded, non-vascular
B. non-seeded, vascular
C. non-seeded, non-vascular
D. seeded, vascular

79. The process in which pollen grains are released from the anthers is called:
(Easy) (Skill 20.2)

A. pollination.
B. fertilization.
C. blooming.
D. dispersal.

80. Water movement to the top of a twenty foot tree is most likely due to which principle?
(Average Rigor) (Skill 20.3)

A. osmostic pressure
B. xylem pressure
C. capillarity
D. transpiration

81. Which of the following is not a strategy used by a young cactus to survive in an arid envrioment?
(Rigorous) (Skill 20.4)

A. Stem as the principle site of photosynthesis.
B. A deep root system to get at sources of groudwater.
C. CAM cycle photosynethsis.
D. Spherical growth form.

82. Characteristics of coelomates include:

    I. no true digestive system
    II. two germ layers
    III. true fluid filled cavity
    IV. three germ layers
    (Average Rigor) (Skill 21.1)

    A. I
    B. II and IV
    C. IV
    D. III and IV

83. Which of the following is a disadvantage of budding as a reproductive mechanism, as compared to sexual reproduction.
    (Rigorous) (Skill 21.2)

    A. limited number of offspring
    B. inefficient
    C. limited genetic diversity
    D. expensive to the parent organism

84. Phylum Mollusca gives a chance to examine both open and closed circulatory systems, with most species having an open system and the cephalopods having a closed system. Which of the following is not shared by open and closed cirulatory systems in molluscs.
    (Rigorous) (Skill 21.3)

    A. Hemocoel
    B. Plasma
    C. Vessels
    D. Heart

85. After sea turtles are hatched on the beach, they start the journey to the ocean. This is due to:
    (Easy) (Skill 21.5)

    A. innate behavior
    B. territoriality
    C. the tide
    D. learned behavior

86. Which of the following answers is best described by the following conditions: not under conscious control and able to maintain a long term contraction.
    (Rigorous) (Skill 22.1)

    A. Fast twitch sketetal muscle
    B. Slow twitch sketetal muscle
    C. Smooth muscle
    D. Cardiac muscle

87. Which of the following compounds is not needed for skeletal muscle contraction to occur?
    (Rigorous) (Skill 22.2)

    A. glucose
    B. sodium
    C. acetylcholine
    D. adenosine 5'-triphosphate

88. Movement is possible by the action of muscles pulling on
(Average Rigor) (Skill 22.3)

A. skin.
B. bones.
C. joints.
D. ligaments.

89. All of the following are found in the dermis layer of skin except:
(Easy) (Skill 22.4)

A. sweat glands.
B. keratin.
C. hair follicles.
D. blood vessels.

90. Capillaries come into contact with a very large surface on both the kidneys and the lungs, especially in relation to the volume of these organs. Which of the following is not consistent with both organs and their contact with capillaries.
(Rigorous) (Skill 23.1)

A. Small specilized sections of each organ
B. A large branching system of tubes within the organ
C. A large source of blood that is quickly divided into capillaries
D. A sack that contains a capillary network

91. What is the principle driving mechanism of inhalation and exhalation?
(Easy) (Skill 23.2)

A. diaphragm
B. the muscles surrounding the ribcage
C. the lungs
D. the bronchial tubes

92. Which of the following parts of the kidney is responsible for the filtering of salt from the body?
(Average Rigor) (Skill 23.3)

A. proximal convoluted tubule
B. the loop of Henle
C. the distal tubule
D. Bowman's capsule

93. Which of the following is not generally a symptom of nephritis?
(Average Rigor) (Skill 23.4)

A. Hypertension
B. Hematuria
C. Edema
D. Necrosis

94. In which of the following blood vessels is blood pressure greatest?
(Easy) (Skill 24.2)

A. veins
B. venules
C. capillaries
D. arteries

95. Congestive heart failure is a devastating disease which, depending on the stage in which it is diagnosed, can be fatal in days or even hours. Which of the following is most the likely treatment to be prescribed by doctors in the beginning stages?

   I   Surgery
   II   Chemotherapy
   III   Diet changes
   IV   Diuretics
   (Rigorous) (Skill 24.3)

   A. I, II
   B. I, III
   C. I, II, III, IV
   D. III, IV

96. A school age boy had chicken pox as a baby. He will most likely not get this disease again because of *(Average Rigor) (Skill 24.4)*

   A. passive immunity
   B. vaccination.
   C. antibiotics.
   D. active immunity.

97. In which of the following situations would you expect doctors to use immunosuppresants? *(Rigorous) (Skill 24.5)*

   A. a skin graft
   B. a blood transfusion
   C. an organ transplant
   D. severe allergic reactions

98. Which of the following substances in unlikely to cause negative consequences if over-ingested? *(Rigorous) (Skill 25.1)*

   A. essenential fatty acids
   B. essenential minerals
   C. essenential water-insoluable vitamins
   D. essenential water-soluable vitamins

99. A muscular adaptation to move food through the digestive system is called: *(Easy) (Skill 25.2)*

   A. peristalsis.
   B. passive transport.
   C. voluntary action.
   D. bulk transport.

100. Fats are broken down by which substance? *(Average Rigor) (Skill 25.3)*

   A. bile produced in the gall bladder
   B. lipase produced in the gall bladder
   C. glucagons produced in the liver
   D. bile produced in the liver

101. Which of the following is not a common gastrointestinal disease?
(Average Rigor) (Skill 25.4)

A. ulcers
B. malabsorption
C. diverticulitis
D. diabetic nephropathy

102. The role of neurotransmitters in nerve action is:
(Average Rigor) (Skill 26.1)

A. to turn off the sodium pump
B. to turn off the calcium pump
C. to send impulses to neurons
D. to send impulses to the body

103. Hormones are essential in the regulation of reproduction. What organ is responsible for the release of hormones for sexual maturity?
(Average Rigor) (Skill 26.2)

A. pituitary gland
B. hypothalamus
C. pancreas
D. thyroid gland

104. Organophosphates, which are used as insecticides, herbcides, and nerve gas, inhibit the breakdown of acetylcholine. This serves as a rather extreme example of which type of drug's action?
(Rigorous) (Skill 26.3)

A. Psychedelic drugs
B. Stimulants
C. Depressants
D. Narcotics

105. In the case of someone experiencing unexplained changes in body tempature, sudden changes in level of hunger, and a change in circadian rythyms, which structure is most likely to be the cause of these problems?
(Rigorous) (Skill 26.5)

A. hypothalamus
B. central nervous system
C. pineal gland
D. basal ganglia

106. Which of the following has no relation to female sexual maturity?
(Rigorous) (Skill 27.1)

A. thyroxine
B. estrogen
C. testosterone
D. luteinizing hormone

107. Fertilization in humans usually occurs in the:
(Easy) (Skill 27.2)

A. cervix
B. ovary
C. fallopian tubes
D. vagina

108. There are many dangers inherent in drug and alcohol abuse, not the least of which is that such abuse can harm a developing fetus. Which of the following might we be surprised to see in the child of an addicted mother?
(Average Rigor) (Skill 27.3)

A. cerebral infarction
B. micropthalmia
C. microcephaly
D. Klinefelter's Syndrome

109. Which of the following best explains the fact that environmental factors are more often the cause of male infertility than female infertility?
(Average Rigor) (Skill 27.4)

A. Women don't work in outdoor professions as frequently as men do, thus limiting their exposure to environmental toxins.
B. Men don't bathe as often as women do, leaving toxins on their hair and skin.
C. Men produce sperm throughout their lives, while women produce no new eggs after puberty.
D. The external nature of the male reproductive organs places them more at risk of exposure to environmental toxins.

110. All of the following are density dependent factors that affect a population except
(Rigorous) (Skill 28.1)

A. disease
B. drought
C. predation
D. migration

111. In the growth of a population, the increase is exponential until carrying capacity is reached. This is represented by a(n):
(Average Rigor) (Skill 28.2)

A. S curve.
B. J curve.
C. M curve.
D. L curve.

112. A clownfish is protected by the sea anemone's tentacles. In turn, the anemone receives uneaten food from the clownfish. This is an example of
(Easy) (Skill 28.3)

A. mutualism.
B. parasitism.
C. commensalism.
D. competition.

113. Population densities are usually estimated using representative plots. If we use a 100 square mile portion of upstate New York as our representative plot, what disperal pattern would you expect?
(Rigorous) (Skill 28.4)

A. random
B. staggered
C. clumped
D. uniform

114. If DDT were present in an ecosystem, which of the following organisms would have the highest concentration in it's system?
(Average Rigor) (Skill 29.1)

A. grasshopper
B. eagle
C. frog
D. crabgrass

115. Which trophic level has the highest ecological efficiency?
(Average Rigor) (Skill 29.2)

A. decomposers
B. producers
C. tertiary consumers
D. secondary consumers

116. High humidity and temperature stability are present in which of the following biomes?
(Easy) (Skill 29.3)

A. taiga
B. deciduous forest
C. desert
D. tropical rain forest

117. Which of the following terms does not describe a way that the human race has had a negative impact on the biosphere?
(Rigorous) (Skill 29.4)

A. Biological magnification
B. Pollution
C. Carrying Capacity
D. Simplifcation of the food web

118. Which term is not associated with the water cycle?
(Easy) (Skill 30.1)

A. precipitation
B. transpiration
C. fixation
D. evaporation

119. In the nitrogen cycle, decomposers are responsible for which process?
(Rigorous) (Skill 30.2)

A. Nitrogen fixing
B. Nitrification
C. Ammonification
D. Assimilation.

120. Which biogeochemical cycle plays the least part in photosynthesis or cellular respiration?
(Rigorous) (Skill 30.3)

A. Hydrogen Cycle
B. Phosphorous Cycle
C. Sulphur Cycle
D. Nitrogen Cycle

121. The concept that the rate of a given process is controlled by the scarcest factor in the process is known as?
(Average Rigor) (Skill 30.4)

A. The Rate of Origination.
B. The Law of the Minimum.
C. The Law of Limitation
D. The Law of Conservation

122. The biggest problem that humans have created in our environment is:
(Average Rigor) (Skill 31.1)

A. Nutrient depletion and excess.
B. Global Warming.
C. Population overgrowth.
D. Ozone depletion.

123. The three main concerns in nonrenewable resource management are conservation, environmental mitigation, and _____.
(Rigorous) (Skill 31.2)

A. Preservation
B. Extraction
C. Allocation
D. Sustainability

124. Which of the following is/are reason/s to maintain biological diversity?

    I. Consumer product development.
    II. Stability of the environment.
    III. Habitability of our planet.
    IV. Cultural Diversity.
    (Rigorous) (Skill 31.3)

    A. I and III
    B. II and III
    C. I, II, and III
    D. I, II, III, and IV

125. Stewardship is the responsible management of resources. We must regulate our actions to do which of the following about environmental degredation?
    (Average Rigor) (Skill 31.5)

    A. Prevent it.
    B. Reduce it.
    C. Mitigate it.
    D. All of the above.

Answer Key

1.	D	26.	C	51.	B	76.	D	101.	D
2.	D	27.	C	52.	B	77.	A	102.	A
3.	D	28.	A	53.	B	78.	B	103.	B
4.	B	29.	A	54.	B	79.	A	104.	B
5.	D	30.	B	55.	A	80.	D	105.	A
6.	D	31.	B	56.	C	81.	B	106.	A
7.	D	32.	D	57.	D	82.	D	107.	C
8.	D	33.	C	58.	D	83.	C	108.	D
9.	C	34.	C	59.	B	84.	A	109.	C
10.	D	35.	B	60.	A	85.	A	110.	B
11.	A	36.	A	61.	D	86.	C	111.	A
12.	C	37.	B	62.	A	87.	A	112.	A
13.	D	38.	C	63.	D	88.	B	113.	C
14.	D	39.	C	64.	C	89.	B	114.	B
15.	C	40.	D	65.	C	90.	D	115.	B
16.	C	41.	B	66.	C	91.	A	116.	D
17.	C	42.	D	67.	C	92.	B	117.	C
18.	C	43.	C	68.	C	93.	D	118.	C
19.	B	44.	C	69.	C	94.	D	119.	C
20.	A	45.	A	70.	A	95.	D	120.	C
21.	D	46.	A	71.	A	96.	D	121.	B
22.	D	47.	C	72.	A	97.	C	122.	A
23.	D	48.	B	73.	D	98.	D	123.	C
24.	A	49.	B	74.	C	99.	A	124.	D
25.	C	50.	C	75.	A	100.	D	125.	D

Rigor Table

	Easy %20	Average Rigor %40	Rigorous %40
Question #	4, 7, 9, 23, 24, 27, 30, 36, 39, 47, 51, 57, 60, 65, 69, 79, 85, 89, 91, 94, 99, 107, 112, 116, 118	3, 5, 10, 12, 13, 15, 18, 19, 20, 26, 28, 34, 35, 37, 40, 41, 44, 45, 48, 49, 50, 54, 56, 59, 67, 65, 66, 70, 72, 73, 74, 76, 80, 82, 88, 92, 93, 96, 100, 101, 102, 103, 108, 109, 11, 114, 115, 121, 122, 125	1, 2, 6, 8, 11, 14, 16, 17, 21, 22, 25, 29, 31, 32, 33, 38, 42, 43, 46, 52, 53, 55, 58, 62, 63, 67, 68, 71, 75, 77, 78, 81, 83, 84, 86, 87, 90, 95, 97, 98, 104, 105, 106, 110, 113, 117, 119, 120, 123, 124

Rationales with Sample Questions

1. One common thing found in math, science, and technology is equilibrium. Which of the following is not an example of dynamic equilibrium?
    (Rigorous) (Skill 1.1)

    A a stable population
    B a symbiotic pair of organisms
    C osmoregulation
    D maintaining head position while walking

Answer: D. maintaining head position while walking
Maintinaing head position while walking is a case of static equilibrium, a state where things do not change. In a stable population birth and death rates must balance. In a symbiotic pair, the contributions of each organism must balance, or the relationship becomes parasistic. Osmoregulation balances the body's need for water, and the body's need to deal with the disolved compounds within it.

2. Which of the following is not a commonly used cross-discipline concept?
    (Rigorous) (Skill 1.2)

    A algebraic substitution of variables
    B data manipulation
    C graphical interpretation of data
    D diversity

Answer: D. Diversity
Algebraic substitution of variables, data manipulation, and graphical interpretation of data are all commonly used in multiple disciplines. Diversity, however, is usually only as a concept used in the social sciences, or in the case of biological diversity, in biology. The answer is (D).

3.   Which of the following limit the development of technological design and solutions?

I. monetary cost
II. time
III. laws of nature
IV. governmental regulation
(Average Rigor) (Skill 1.3)

A   I and II
B   I, II, and IV
C   II and III
D   I, II, and III

Answer: D. I, II, and III
Technology can not work against the laws of nature. Technological design solutions must work within the framework of the natural world. Monetary cost and time constraints also limit the development of new technologies. Governmental regulation, while present in many sciences, cannot regulate the formation of new ideas or design solutions. The answer, then, is D: I, II, and III.

4.   Computer-linked probes automatically gather data and present it in an accessible format. This eliminates the need for:
(Easy) (Skill 1.4)

A   scientific inquiry
B   constant human observation
C   data analysis
D   experimentation

Answer: B. Constant human observation
Scientific Inquiry, Data Analysis, and Experimentation are all concepts that require human activity to complete, but computer-linked probes can eliminate the need for constant human observation, as they can monitor temperature, movement, and activity and make these observations accessible. The answer is (B).

5.   The individual parts of cells are best studied using a(n)...
     (Average Rigor) (Skill 2.1)

     A   ultracentrifuge.
     B   phase-contrast microscope.
     C   CAT scan.
     D   electron microscope.

Answer: D. electron microscope.
The scanning electron microscope uses a beam of electrons to pass through the specimen. The resolution is about 1000 times greater than that of a light microscope. This allows the scientist to view extremely small objects, such as the individual parts of a cell.

6.   Who discovered penicillin?
     (Rigorous) (Skill 2.2)

     A   Pierre Curie
     B   Becquerel
     C   Louis Pasteur
     D   Alexander Fleming

Answer: D. Alexander Fleming
Sir Alexander Fleming was a pharmacologist from Scotland. He isolated the antibiotic penicillin from a fungus in 1928. Pierre Curie, Marie Curie, and Becquerel are most known for their work with Radium. Louis Pasteur is known for his work in mibrobiology; he created the first rabies vaccine.

7.   The demand for genetically enhanced crops has increased in recent years. Which of the following is not a reason for this increased demand?
     (Easy) (Skill 2.3)

     A   Fuel sources
     B   Increased growth
     C   Insect resistance
     D   Better-looking produce

Answer: D. Better-Looking Produce
Genetically enhanced crops are being developed for utilization as fuel sources, as well as for an increased production yield. Insect resistance is being bred into crops to eliminate the need for pesticides. While there are surely some farmers out there crossing crops to make prettier watermelons, this is not a primary reason for the increased demand. The answer is (D).

8.   **The foremost technique for isolating genes is:**
(Rigorous) (Skill 2.4)

A   restrictive enzyme digestion
B   polymerized chain reaction
C   northern blot
D   cDNA cloning

Answer: D. cDNA cloning
Restrictive enzyme digestion is used to cut up DNA; Polymerized chain reaction is used to duplicate DNA you have already obtained; Northern Blot is an older technique used to isolate DNA fragments. cDNA cloning technique starts with RNA and works backwards, a simpler process than many of the others. cDNA is the most widely-used gene isolation technique. The answer is (D).

9.   **Which is the correct order of methodology?**

1) testing a revised explanation,
2) setting up a controlled experiment to test an explanation,
3) drawing a conclusion,
4) suggesting an explanation for observations,
5) comparing observed results to hypothesized results
(Easy) (Skill 3.1)

A   4, 2, 3, 1, 5
B   3, 1, 4, 2, 5
C   4, 2, 5, 1, 3
D   2, 5, 4, 1, 3

Answer: C. 4, 2, 5, 1, 3
The first step in scientific inquiry is posing a question to be answered. Next, a hypothesis is formed to provide a plausible explanation. An experiment is then proposed and performed to test this hypothesis. A comparison between the predicted and observed results is the next step. Conclusions are then formed and it is determined whether the hypothesis is correct or incorrect. If incorrect, the next step is to form a new hypothesis and repeat the process.

10. **The idea that all rational beings have intrinsic value is the backbone of:**
    (Average Rigor) (Skill 3.2)

    A   Utilitarianism
    B   Social Contract Theory
    C   Darwinism
    D   Kantianism

Answer: D. Kantianism
Philosophers have given a number of moral theories to justify moral rules, which range from utilitarianism (a theory of ethics that prescribes the quantitative maximization of good consequences for a population proposed by Mozi, a Chinese philosopher who lived from 471-381 BC, Kantianism (a theory proposed by Immanuel Kant, a German philosopher who lived from 1724-1804, which ascribes intrinsic value to rational beings and is the philosophical foundation of contemporary human rights) to social contract theory (a view of the ancient Greeks which states that the person's moral and or political obligations are dependent upon a contract or agreement between them to form society).

11. **As a science fair project, a student cuts a potato into four quarters and plants each in a different type of soil, then administers the same amount of water to each and allows each to receive the same amount of sunlight. Which of the following is the most likely hypothesis being tested?**
    (Rigorous) (Skill 3.3)

    A   The different soil environments would effect the growing conditions more than the genetic uniformity of the samples.
    B   The potatoes will grow at the same rate.
    C   One half (two pieces) of the potato will grow faster than the other half.
    D   No growth will occur in the majority of the soils.

Answer: A. The different soil environments would effect the growing conditions more than the genetic uniformity of the samples.
One can assume that the student wouldn't be doing the experiment if they expected the potatoes to grow at the same rate, or if we expected no growth to occur. The idea that half of the potato would grow and the other would not is a rather arbitrary one. That the different soil environments affect the growing conditions is the best example of an educated guess. The answer is (A).

12.   **A student designed a science project testing the effects of light and water on plant growth. You would recommend that the student:**
      (Average Rigor) (Skill 3.4)

    A   manipulate the temperature as well.
    B   also alter the pH of the water as another variable
    C   omit either water or light as a variable
    D   also alter the light concentration as another variable.

Answer: C. omit either water or light as a variable
In science, experiments should be designed so that only one variable is manipulated at a time.

13.   **A scientific theory:**
      (Average Rigor) (Skill 4.1)

    A   proves scientific accuracy.
    B   is never rejected.
    C   results in a medical breakthrough.
    D   may be altered at a later time.

Answer: D. may be altered at a later time.
Scientific theory is usually accepted and verified information but can always be changed at anytime.

14.   **Which graphical representation would best compare the rate of activity of different enzymes at various temperatures?**
      (Rigorous) (Skill 4.2)

    A   Pie Chart
    B   Histogram
    C   Bar Graph
    D   Line Graph

Answer: D. Line Graph
Line graphs compare different sets of related data and help predict data. A bar graph or histogram compares different items and helps make comparisons based on the data. A pie chart is useful when organizing data as part of a whole. The correct answer is (D).

15. **In a list of data, the value of the item that occurs the most often is:**
     (Average Rigor) (Skill 4.4)

     A   The Mean
     B   The Median
     C   The Mode
     D   The Range

Answer: C. The Mode
Mean is the mathematical average of all the items. The median depends on whether the number of items is odd or even. If the number is odd, then the median is the value of the item in the middle. Mode is the value of the item that occurs the most often, if there are not many items. Bimodal is a situation where there are two items with equal frequency. Range is the difference between the maximum and minimum values.

16. **In which of the following situations would a linear extrapolation of data be appropriate?**
     (Rigorous) (Skill 4.5)

     A   Computing the death rate of an emerging disease
     B   Computing the number of plant species in a forest over time
     C   Computing the rate of diffusion with a constant gradient
     D   Computing the population at equilibrium

Answer: C: Computing the rate of diffusion with a constant gradient.
The individual data points on the graph of a linear relationship cluster around a line of best fit. In other words, a relationship is linear if we can sketch a straight line that roughly fits the data points. Extrapolation is the process of estimating data points outside a known set of data points. When extrapolating data of a linear relationship, we extend the line of best fit beyond the known values. The extension of the line represents the estimated data points. Extrapolating data is only appropriate if we are relatively certain that the relationship is indeed linear. The answer is (C).

17. **The SI measurement for temperature is on the _____ scale.**
    (Rigorous) (Skill 5.1)

    A   Celsius
    B   Farenheit
    C   Kelvin
    D   Rankine

Answer: C. Kelvin
Science uses the metric system (the SI system) because of its worldwide acceptance and ease of comparison. The SI scale for measuring temperature is the Kelvin Scale. Science, however, uses the Celsius scale for its ease of use. The answer is (C).

18. **The reading of a meniscus in a graduated cylinder is done at the...**
    (Average Rigor) (Skill 5.2)

    A   top of the meniscus.
    B   middle of the meniscus.
    C   bottom of the meniscus.
    D   closest whole number.

Answer: C. bottom of the meniscus.
The graduated cylinder is the common instrument used for measuring volume. It is important for the accuracy of the measurement to read the volume level of the liquid at the bottom of the meniscus. The meniscus is the curved surface of the liquid.

19. **Electrophoresis separates DNA on the basis of...**
    (Average Rigor) (Skill 6.1)

    A   amount of current.
    B   molecular size.
    C   positive charge of the molecule.
    D   solubility of the gel.

Answer: B. molecular size.
Electrophoresis uses electrical charges of molecules to separate them according to their size.

20.   The "Right to Know" law states:
       (Average Rigor) (Skill 6.2)

    A   that the inventory of toxic chemicals checked against the
            "Substance List" be available.
    B   that students are to be informed of alternatives to dissection.
    C   that science teachers are to be informed of student allergies.
    D   that students are to be informed of infectious microorganisms used
            in lab.

**Answer: A. that the inventory of toxic chemicals checked against the
             "Substance List" be available.**
The right to know law pertains to chemical substances in the lab. Employees
should check the material safety data sheets and the substance list for potential
hazards in the lab.

21.   Which of the following is not usually found on the MSDS for a
       laboratory chemical?
       (Rigorous) (Skill 6.3)

    A   Melting Point
    B   Toxicity
    C   Storage Instructions
    D   Cost

Answer: D. Cost
MSDS, or Materials Safety Data Sheets, are used to make sure that anyone
concerned can easily obtain information about the chemical used in the event of
a spill or accident. This information typically includes physical data, toxicity,
health effects, first aid, reactivity, storage, disposal, protective measures, and
spill/leak procedures. Cost is not generally included as MSDS's are produced by
the distributor, and seperate suppliers may have different costs. The answer,
therefore, is (D).

22.   **Which of the following is the least ethical choice for a school laboratory activity?**
      (Rigorous) (Skill 6.4)

    A   Dissection of a donated cadaver.
    B   Dissection of a preserved fetal pig.
    C   Measuring the skeletal remains of birds.
    D   Pithing a frog to watch the circulatory system.

Answer: D. Pithing a frog to watch the circulatory system.
Scientific and societal ethics make choosing experiments in today's science classroom difficult. It is possible to ethically perform choices (A), (B), or (C), if due care is taken. (Note that students will need significant assistance and maturity to perform these experiments, and that due care also means attending to all legal paperwork that might be necessary.) However, modern practice precludes pithing animals (causing partial brain death while allowing some systems to function), as inhumane. Therefore, the answer to this question is (D).

23.   **Who should be notified in the case of a serious chemical spill?**

    I.   **the custodian**
    II.   **the fire department**
    III. **the chemistry teacher**
    IV. **the administration**
    (Easy) (Skill 6.5)

    A   I
    B   II
    C   II and III
    D   II and IV

Answer: D. II and IV
For large spills, the school administration and the local fire department should be notified.

24. **Which kingdom is comprised of organisms made of one cell with no nuclear membrane?**
    (Easy) (Skill 7.1)

    A   Monera
    B   Protista
    C   Fungi
    D   Algae

Answer: A. Monera
Monera is the only kingdom that is made up of unicellular organisms with no nucleus. Algae is a protist because it is made up of one type of tissue and it has a nucleus.

25. **Which of the following is not considered evidence of Endosymbiotic Theory?**
    (Rigorous) (Skill 7.2)

    A   The presence of genetic material in mitochondria and plastids.
    B   The presence of ribosomes within mitochondria and plastids.
    C   The presence of a double layered membrane in mitochondria and plastids.
    D   The ability of mitochondria and plastids to reproduce.

Answer: C. The presence of a double layered membrane in mitochondria and plastids.
A double layered membrane is not unique to mitochondria and plastids, a cell's nucleus is also double layered. All three other characteristics are not present in anyother organelles, and are evidence that thse two organelles once may have been seperate organisms.

26. **What is necessary for diffusion to occur?**
    (Average Rigor) (Skill 7.3)

    A   carrier proteins
    B   energy
    C   a concentration gradient
    D   a membrane

Answer: C. a concentration gradient
Diffusion is the ability of molecules to move from areas of high concentration to areas of low concentration (a concentration gradient).

27.   **Negatively charged particles that circle the nucleus of an atom are called...**
      (Easy) (Skill 8.1)

      A   neutrons.
      B   neutrinos.
      C   electrons.
      D   protons.

Answer: C. electrons.
Neutrons and protons make up the core of an atom. Neutrons have no charge and protons are positively charged. Electrons are the negatively charged particles around the nucleus.

28.   **Which protein structure consists of the coils and folds of polypeptide chains?**
      (Average Rigor) (Skill 8.2)

      A   secondary structure
      B   quaternary structure
      C   tertiary structure
      D   primary structure

Answer: A. secondary structure
Primary structure is the protein's unique sequence of amino acids. Secondary structure is the coils and folds of polypeptide chains. The coils and folds are the result of hydrogen bonds along the polypeptide backbone. Tertiary structure is formed by bonding between the side chains of the amino acids. Quaternary structure is the overall structure of the protein from the aggregation of two or more polypeptide chains.

29.   **Which of the following are properties of water?**

    I.   **High specific heat**
    II.   **Strong hydrogen bonds**
    III. **Good solvent**
    IV. **High freezing point**
(Rigorous) (Skill 8.3)

    A   I, III, IV
    B   II and III
    C   I and II
    D   II, III, IV

Answer: A. I, III, IV
All are properties of water except strong hydrogen bonds. Water is held together by polar covalent bonds between hydrogen and oxygen. In a liquid state hydrogen bonding is the cause of water's cohesive properties, however the individual hydrogen bonds are easily broken.

30.   **Sulfur oxides and nitrogen oxides in the environment react with water to cause:**
(Easy) (Skill 8.4)

    A   ammonia.
    B   acidic precipitation.
    C   sulfuric acid.
    D   global warming.

Answer: B. acidic precipitation.
Acidic precipitation is rain, snow, or fog with a pH less than 5.6. It is caused by sulfur oxides and nitrogen oxides that react with water in the air to form acids that fall down to Earth as precipitation.

31. **ATP is known to bind to phosphofructokinase-1 (an enzyme that is part of glycolysis). This results in a change in the shape of the enzyme and causes the rate of ATP production to fall. Which answer best explores what has happened?**
    (Rigorous) (Skill 8.5)

    A   binding of a coenzyme
    B   an allosteric change in the enzyme
    C   competitive inhibition
    D   uncompetive inhibition

Answer: B. an allosteric change in the enzyme
The binding of ATP to phosphofructokinase-1 causes an allosteric change (a change in shape) of the enzyme. The binding of ATP can be considered non-competitive inhibition.

32. **During the Krebs cycle, 8 carrier molecules are formed. What are they?**
    (Rigorous) (Skill 9.2)

    A   3 NADH, 3 FADH$_2$, 2 ATP
    B   6 NADH and 2 ATP
    C   4 FADH$_2$ and 4 ATP
    D   6 NADH and 2 FADH$_2$

Answer: D. 6 NADH and 2 FADH$_2$
For each molecule of CoA that enters the Krebs cycle, you get 3 NADH and 1 FADH$_2$. There are 2 molecules of CoA so the total yield is 6 NADH and 2 FADH$_2$ during the Krebs cycle.

33. **The most ATP is generated through...**
    (Rigorous) (Skill 9.3)

    A   fermentation.
    B   glycolysis.
    C   chemiosmosis.
    D   the Krebs cycle.

Answer: C. chemiosmosis.
The electron transport chain uses electrons to pump hydrogen ions across the mitochondrial membrane. This ion gradient is used to form ATP in a process called chemiosmosis. ATP is generated by the removal of hydrogen ions from NADH and FADH$_2$. This yields 34 ATP molecules.

34. **Which of the following is not a factor that affects the rate of both photosynthesis and respiration in plants?**
(Average Rigor) (Skill 9.4)

    A   the concentration of NADP and FAD
    B   the temperature
    C   the structure of the plants
    D   the availability of the different substrates

Answer: C. the structure of the plants
The structure of the plants's leaf affects its ability to absorb light which then affects the rate of photosynthesis. Although both processsess require different substrates they are both dependent on the concentration of those substrates (e.g. carbon dioxide and oxygen).

35. **The product of anaerobic respiration in animals is:**
(Average Rigor) (Skill 9.5)

    A   carbon dioxide.
    B   lactic acid.
    C   pyruvate.
    D   ethyl alcohol.

Answer: B. lactic acid.
In anaerobic lactic acid fermentation, pyruvate is reduced by NADH to form lactic acid. This is an anaerobic process in animals. Alcoholic fermentation is the anaerobic process in yeast and some bacteria resulting in ethyl alcohol. Carbon dioxide and pyruvate are the products of aerobic respiration.

36. **Which of the follow is not true of both chloroplasts and mitochondria?**
(Easy) (Skill 9.6)

    A   use of the inner membrane for most of it's activity
    B   converts energy from one form to another
    C   uses an electron transport chain
    D   is an important part of the carbon cycle

Answer: A. use of the inner membrane for most of it's activity
In mitochondria the electron transport chain is present in the inner membrane, however in chloroplasts it is present in the thylakoid membranes.

37. **DNA synthesis results in a strand that is synthesized continuously. This is the...**
(Average Rigor) (Skill 10.1)

    A   lagging strand
    B   leading strand
    C   template strand
    D   complementary strand

Answer: B. leading strand.
As DNA synthesis proceeds along the replication fork, one strand is replicated continuously (the leading strand) and the other strand is replicated discontinuously (the lagging strand).

38. **Which of the following is not posttranscriptional processing?**
(Rigorous) (Skill 10.2)

    A   5' capping
    B   intron splicing
    C   polypeptide splicing
    D   3' polyadenylation

Answer: C. polypeptide splicing
The removal of segments of polypeptides is a posttranslational process. The other three are methods of posttranscriptional processing.

39. **Any change that affects the sequence of bases in a gene is called a(n):**
(Easy) (Skill 10.3)

A   deletion
B   polyploid
C   mutation
D   duplication

Answer: C. mutation.
A mutation is an inheritable change in DNA. It may be an error in replication or a spontaneous rearrangement of one ore more segments of DNA. Deletion and duplication are types of mutations. Polyploidy is when an organism has more than two complete chromosome sets.

40. **The lac operon**

I. **contains the lac Z, lac Y, and lac A genes**
II. **converts glucose to lactose**
III. **contains a repressor**
IV. **is on when the repressor is activated**
(Average Rigor) (Skill 10.4)

A   I
B   II
C   III and IV
D   I and III

Answer: D. I and III
The lac operon contains the genes that encode for the enzymes used to convert lactose into fuel. It contains three genes: lac A, lac Z, and lac Y. It also contains a promoter and repressor. When the repressor is activated, the operon is off.

41.   **The polymerase chain reaction**
        (Average Rigor) (Skill 11.2)

        A   is a group of polymerases.
        B   is a technique for amplifying DNA.
        C   is a primer for DNA synthesis.
        D   is synthesis of polymerase.

Answer: B. is a technique for amplifying DNA.
PCR is a technique in which a piece of DNA can be amplified into billions of
copies within a few hours.

42.   **Which is not a way that genetic engineering advancement has been
        used to create a useful application.**
        (Rigorous) (Skill 11.3)

        A   The creation of safer viral vacines.
        B   The creation of bacteria that produce hormones for medical use.
        C   The creation of bacteria to breakdown toxic waste.
        D   The creation of organisms that are successfully being used as
                sources for alternative fuels.

**Answer: D. The creation of organisms that are successfully being used as
                sources for alternative fuels.**
Although there is a push for the use of genetic engineering to create organisms
that will either create alternative fuels, be used as an alternative fuel, or increase
the efficiency of those organisms that are already being used as a fuel source
(e.g. biodiesel engines that can run on various plant oils.) This field is just
starting to develop whereas there are multiple successful examples for each of
the other posssible answers.

43.   **Which is a way that cDNA cloning has not been used?**
        (Rigorous) (Skill 11.4)

        A   to provide evidence for taxonomic organization
        B   to study the mutations that lead to diseases such as hemophilia
        C   to determine the structure of a protein
        D   to understand the methods of gene regulation

Answer: C. to determine the structure of a protein
Although cDNA cloning can be used to determine the amino acid sequence of a
protein many other steps determine the final protein structure. For example the
folding of the protein, addition of other protein subunits, and/or modification by
other proteins.

44. **Which of the following is not a way in which the U.S. government has regulated genetic engineering?**
   (Average Rigor) (Skill 11.6)

   A   stem cell research
   B   whether engineered food may go to market
   C   the isolation and study of genes from nonpathogenic organisms
   D   the use of gene therapy treatments

Answer: C. the isolation and study of genes from nonpathogenic organisms

Although there may be limits placed on research by a funding agency, at this time there are no federal regulations on the isolation and study of genes from nonpathogenic organisms.

45.   **Identify this stage of mitosis.**

   (Average Rigor) (Skill 12.1)

   A   anaphase
   B   metaphase
   C   prophase
   D   telophase

Answer: A. Anaphase

During anaphase, the centromeres split in half and homologous chromosomes separate.

46.   **Which is not a possible effect of polyploidy?**
        (Rigorous) (Skill 12.2)

        A   more robust members of a species
        B   the creation of a cross species offspring
        C   the creation of a new species
        D   cells that produce higher levels of desired proteins

Answer: A. more robust members of a species
While polyploidy often creates more robust crops, it is likely to create nonviable animal offspring, and in plants usually a new species will develop (there are even a few rare animal examples).

47.   **Which process(es) result(s) in a haploid chromosome number?**
        (Easy) (Skill 12.3)

        A   both meiosis and mitosis
        B   mitosis
        C   meiosis
        D   replication and division

Answer: C. Meiosis
In meiosis, there are two consecutive cell divisions resulting in the reduction of the chromosome number by half (diploid to haploid).

48.   **Cancer cells divide extensively and invade other tissues. This continuous cell division is due to:**
        (Average Rigor) (Skill 12.5)

        A   density dependent inhibition.
        B   density independent inhibition.
        C   chromosome replication.
        D   inhibition hormones.

Answer: B. density independent inhibition.
Density dependent inhibition is when the cells crowd one another and consume all the nutrients, thereby halting cell division. Cancer cells, however, are density independent; meaning they can divide continuously as long as nutrients are present.

49. **The ratio of brown-eyed to blue-eyed children from the mating of a blue-eyed male to a heterozygous brown-eyed female would be expected to be which of the following?**
(Average Rigor) (Skill 13.1)

A   2:1
B   1:1
C   1:0
D   1:2

Answer: B. Use a Punnet square to determine the ratio.

	b	b
B	Bb	Bb
b	bb	bb

B = brown eyes, b = blue eyes

Female genotype is on the side and the male genotype is across the top.

The female is heterozygous and her phenotype is brown eyes. This means the dominant allele is for brown eyes. The male expresses the homozygous recessive allele for blue eyes. Their children are expected to have a ratio of brown eyes to blue eyes of 2:2; or 1:1.

50. **Amniocentesis is:**
(Average Rigor) (Skill 13.2)

A   a non-invasive technique for detecting genetic disorders.
B   a bacterial infection.
C   extraction of amniotic fluid.
D   removal of fetal tissue.

Answer: C. extraction of amniotic fluid.
Amniocentesis is a procedure in which a needle is inserted into the uterus to extract some of the amniotic fluid surrounding the fetus. Some genetic disorders, such as Down's Syndrome, can be detected by analyzing samples of the fluid.

51.   **Sutton observed that genes and chromosomes behave the same. This led him to his theory which stated**
      (Easy) (Skill 13.3)

      A   that meiosis causes chromosome separation.
      B   that linked genes are able to separate.
      C   that genes and chromosomes have the same function.
      D   that genes are found on chromosomes.

Answer: B. that linked genes are able to separate.
Sutton observed how mitosis and meiosis confirmed Mendel's theory on "factors". His Chromosome Theory states that genes are located on chromosomes.

52.

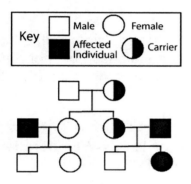

      Based on the pedigree chart, what term best describes the nature of the trait being mapped?
      (Rigorous) (Skill 13.4)

      A   autosomal recessive
      B   sex-linked
      C   incomplete dominance
      D   co-dominance

Answer: B. sex-linked
This chart would be a good example of color blindness, a sex-linked trait. If the trait had been autosomal recessive the last generation would all be carriers with the exception of the affected individual. In cases of traits that are incomplete dominace and co-dominace the tree would require additional notations.

53. **A women has Pearson Syndrome, a disease caused by a mutation in the mitochondrial DNA. Which of the following individuals would you expect to see the disease in?**

I   Her Daughter
II   Her Son
III Her Daughter's son
IV   Her Son's daughter
(Rigorous) (Skill 13.5)

A   I, III
B   I, II, III
C   II, IV
D   I,II, III, IV

Answer: B. I, II, III
Since the mitochondrial DNA is passed through the maternal line, both of her children would be affected and the trait would continue to pass from her daughter to all of her children. Her son's children would receive their mitochondrial DNA from their mother.

54. **Which of the following factors will affect the Hardy-Weinberg law of equilibrium, leading to evolutionary change?**
(Average Rigor) (Skill 14.1)

A   no mutations
B   non-random mating
C   no immigration or emigration
D   large population

Answer: B. non-random mating
There are five requirements to keep the Hardy-Weinberg equilibrium stable: no mutation, no selection pressures, an isolated population, a large population, and random mating.

55. **Which of the following is an example of a phenotype that gives the organism an advantage in their home environment?**
(Rigorous) (Skill 14.2)

    A   The color of Pepper Moths in England
    B   Thornless roses in a nature perserve
    C   Albinoism in naked mole rats
    D   Mediterranean fruit fly having a large thorax

Answer: A. The color of Pepper Moths in England
Thornless roses are not naturally occuring and would not convey an advantage against a rose's natural predators. Albinoism is not any more common in naked mole rats than most species and would convey no advantage in a subterranean environment. Mediterranean fruit fly thorax size has been linked to sexual selection, however sexual selection is not an environmental pressure. The Pepper Moth of England is often used as an example of natural selection in action. A dramatic shift in color frequency occured during the industrial revolution. This is attributed to soot changing the color of trees and resulted in a corresponding change in the color of the moths.

56. **All of the following gasses made up the primitive atmosphere except...**
(Average Rigor) (Skill 15.1)

    A   ammonia
    B   methane
    C   oxygen
    D   hydrogen

Answer: C. Oxygen
In the 1920's, Oparin and Haldane were to first to theorize that the primitive atmosphere was a reducing atmosphere without oxygen. The amosphere was rich in hydrogen, methane, water, and ammonia.

57.   The first cells that evolved on Earth were probably of which type?
        (Easy) (Skill 15.2)

        A   autotrophs
        B   eukaryotes
        C   heterotrophs
        D   prokaryotes

Answer: D. Prokaryotes
Prokaryotes date back to 3.5 billion years ago in the fossil record. Their ability to adapt to the environment allows them to thrive in a wide variety of habitats.

58.   **Fossils of a dinosaur genus known as Saurolophus have been found in both Western North America and in Mongolia. What is the most likely explanation for these findings?**
        (Rigorous) (Skill 15.3)

        A   Convergent evolution on the two continents lead to two species of dinosaurs with significant enough similarities to be placed in the same genus.
        B   With the gaps in the fossil record there are currently mutliple competing theories to explain the presence of these fossils on two separate continents.
        C   A distant ancestor of the these dinosaurs evolved at a point before these land masses became separated.
        D   Although Asia and North America were separate continents at the time, low sea levels made it possible for the dinosaurs to walk from one continent to the other.

Answer: D. Although Asia and North America were separate continents at the time, low sea levels made it possible for the dinosaurs to walk from one continent to the other.
Convergent evolution can explain how different species have developed similar traits but these species would end up each in a different genus because of their greater differences. In this case the fossil record indicates too many similarities, and thus the Saurolophus are a single genus of dinosaurs. Although there are gaps in the fossil record, there is one theory that is supported by significant evidence and to which most of the paleontologists ascribe. The last time these two land masses were part of the same continent (known as Laurasi) was about 40 million years before the Saurolophus was alive, so it is unlikely that evolution would have created a division that would split the genus. This leaves the ability to walk from Asia to North America, and this was accomplished by means of the Bering Land Bridge (where the Bering Strait is now). The Bering Land bridge existed because the level of the water in the oceans was lowered due to water being stored in large glaciers. This lowering of ocean level was enough to expose what is now the ocean floor between Alaska and Siberia.

59.   **Which aspect of science does not support evolution?**
        (Average Rigor) (Skill 15.4)

        A   comparative anatomy
        B   organic chemistry
        C   comparison of DNA among organisms
        D   analogous structures

Answer: B. organic chemistry
Comparative anatomy is the comparison of characteristics of the anatomies of
different species. This includes homologous structures and analogous structures.
The comparison of DNA between species is the best known way to place species
on the evolutionary tree. Organic chemistry has nothing to do with evolution.

60.   **Which process contributes to the large variety of living things in the
        world today?**
        (Easy) (Skill 16.1)

        A   meiosis
        B   asexual reproduction
        C   mitosis
        D   alternation of generations

Answer: A. Meiosis
Crossing over occurs during meiosis prophase I. This exchange of genetic
material between homologues increases diversity.

61.   **An animal choosing its mate because of attractive plumage or a
        strong mating call is an example of:**
        (Average Rigor) (Skill 16.2)

        A   Mechanical isolation.
        B   Natural Selection.
        C   Directional Selection.
        D   Sexual Selection

Answer: D. Sexual Selection
The coming together of genes determines the makeup of the gene pool. Sexual
selection, the act of choosing a mate, allows animals to have some choice in the
breeding of its offspring.

62. **Which of the following best exemplifies the Theory of Inheritance of Acquired Traits?**
    (Rigorous) (Skill 16.3)

    A   Giraffes need to reach higher for leaves to eat, so their necks stretch. The giraffe babies are then born with longer necks. Eventually, there are more long-necked giraffes in the population.

    B   Giraffes with longer necks are able to reach more leaves, so they eat more and have more babies than other giraffes. Eventually, there are more long-necked giraffes in the population.

    C   Giraffes want to reach higher for leaves to eat, so they release enzymes into their bloodstream, which in turn causes fetal development of longer-necked giraffes. Eventually, there are more long-necked giraffes in the population.

    D   Giraffes with long necks are more attractive to other giraffes, so they get the best mating partners and have more babies. Eventually, there are more long-necked giraffes in the population.

Answer: A. Giraffes need to reach higher for leaves to eat, so their necks stretch. The giraffe babies are then born with longer necks. Eventually, there are more long-necked giraffes in the population.

The theory of inheritance of acquired traits states that the offspring of an individual will benefit from the adaptations of the parent. The stretching of the neck thus leads to offspring with longer necks. Answer B best exemplifies the theory of natural selection, where an outside factor is affecting the chance of an individual to live and reproduce, and thus pass on their genetic material to the next generation. There is no evidence of desire creating genetic or developmental change in a fetus. Additionally there is no evidence that giraffes select mates based on neck length, however if they did this would be an example of sexual selection, an aspect of natural selection.

63. The fossil record is often pointed to as evidence of punctuated equilibrium. Which of the following examples can only be explained by punctuated equilibrium and not gradualism?

I Coelacanth fish (once thought extinct) have been relatively unchanged for millions of years.
II The sudden appearance of a large number of different soft-bodied animals around 530 million years ago.
III 10 million year old fossils and modern ginko plants are nearly identical.
IV Fossils of Red Deer for the Island of Jersey show a six-fold decrease in size in body weight in the last 6000 years.

(Rigorous) (Skill 16.4)

A I, III
B II, IV
C I, II, III, IV
D None of the above

Answer: D. None of the above
Gradualism and punctuated equilibrium and not mutually exclusive. Since we are talking in terms of geological time, a rapid change can be thought to occur over a period of 1,000 years to 100,000 years or more. Items I and III appear to demonstrate a static state, however there easily could have been some changes over that time that can not be seen in the fossil record. Items II and IV appear to show a period of sudden change. In the case of item II that change may easily be over the course of a million years or more. In the case of item IV, 6000 years can conceivably include 3000 generations (they are mature at the age of 2). In both cases the apparently sudden change may have actually occurred gradually.

64. The biological species concept applies to
(Average Rigor) (Skill 16.5)

A asexual organisms.
B extinct organisms.
C sexual organisms.
D fossil organisms.

Answer: C. sexual organisms.
The biological species concept states that a species is a reproductive community of populations that occupy a specific niche in nature. It focuses on reproductive isolation of populations as the primary criterion for recognition of species status. The biological species concept does not apply to organisms that are completely asexual in their reproduction, fossil organisms, or distinctive populations that hybridize.

65.   **Members of the same species:**
       (Easy) (Skill 17.1)

    A   look identical.
    B   never change.
    C   reproduce successfully within their group.
    D   live in the same geographic location.

Answer: C. reproduce successfully within their group.
Species are defined by the ability to successfully reproduce with members of their own kind.

66.   **The majority of phyogenetic trees currently created are based upon what?**
       (Average Rigor) (Skill 17.2)

    A   analogous structures
    B   homologous structures
    C   genetic analysis
    D   behavioral trends

Answer: C. genetic analysis
As genetic testing has become cheaper and more efficient it has been increasingly used to reconstruct, augment or replace earlier phylogentic trees that had been based on one or more of the other possible answers.

67.

Using the following taxonomic key identify the tree that this branch came from

1   Are the leaves PALMATELY COMPOUND (BLADES arranged like fingers on a hand)? - 2

1   Are the leaves PINNATELY COMPOUND (BLADES arranged like the veins of a feather)? - 3

2   Are there usually 7 BLADES - Aesculus hippocastanum

2   Are there usually 5 BLADES - Aesculus glabra

3   Are there mostly 3-5 BLADES that are LOBED or coarsely toothed? - Acer negundo

3   Are there mostly 5-13 BLADES with smooth or toothed edges? - Fraxinus americana

(Rigorous) (Skill 17.3)

A   Aesculus hippocastanum
B   Aesculus glabra
C   Acer negundo
D   Fraxinus americana

Answer: C. Acer negundo
The leaves are pinnately compound, with 5 coarsely toothed leaves, leading to the answer: Acer negundo. The list below includes the scientific name and the common name for the all the plants listed above.
Aesculus hippocastanum (Horsechestnut)
Aesculus glabra (Ohio Buckeye)
Acer negundo (Boxelder, Ashleaf Maple)
Fraxinus americana (White Ash)

68. Which of the following systems uses Archea (or Archeabacteria) as the most inclusive level of the taxonomic system.

    I   Three Domain System
    II   Five Kingdom System
    III Six Kingdom System
    IV   Eight Kingdom System
    (Rigorous) (Skill 17.5)

    A   II, III
    B   I, IV
    C   I, III, IV
    D   I, II, III, IV

Answer: C. I, III, IV
In the five kingdom system the subkingdom Archaebacteriobionta is under the kingdom Monera.

69. Identify the correct sequence of organization of living things. *(Easy) (Skill 18.2)*

    A   cell – organelle – organ system – tissue – organ – organism
    B   cell – tissue – organ – organ system – organelle – organism
    C   organelle – cell – tissue – organ – organ system – organism
    D   tissue – organelle – organ – cell – organism – organ system

Answer: C. organelle – cell – tissue – organ – organ system – organism
An organism, such as a human, is comprised of several organ systems such as the circulatory and nervous systems. These organ systems consist of many organs including the heart and the brain. These organs are made of tissue such as cardiac muscle. Tissues are made up of cells, which contain organelles like the mitochondria and Golgi apparatus.

70.   **Which of the following is not a necessary characteristic of living things?**
        (Average Rigor) (Skill 18.3)

        A   Movement
        B   Reduction of local entropy
        C   Ability to cause change in local energy form
        D   Reproduction

Answer: A. Movement
There are many definitions of "life," but in all cases, a living organism reduces local entropy, changes chemical energy into other forms, and reproduces.
Not all living things move, however, so the correct answer is (A).

71.   **Which of the following hormones is most involved in the process of osmoregulation?**
        (Rigorous) (Skill 18.4)

        A   Antidiuretic Hormone
        B   Melatonin
        C   Calcitonin
        D   Gulcagon

Answer: A. Antidiuretic Hormone.
Osmoregulation relates to the body's attempt to control the concentration of water and different soluble materials in the body. Antidiruetic Hormone (ADH) regulates the kidneys' reabsorption of water and so directly relates to the amount of water in the body. The body's failure to produce ADH can lead to a life threatening condition where an individual dies of dehydration within a matter of hours. Calcitonin controls the removal of calcium from the blood. Glucagon, like insulin, controls the amount of glucose in the blood. Melatonin is involved in homeostasis, however it is not directly involved in osmoregulation, rather it is involved in the regulation of body rhythms.

72. **The shape of a cell depends on its...**
   (Average Rigor) (Skill 18.5)

   A   function.
   B   structure.
   C   age.
   D   size.

Answer: A. function.
In most living organisms, cellular structure is based on function.

73. **A virus that can remain dormant until a certain environmental condition causes its rapid increase is said to be:**
   (Average Rigor) (Skill 19.1)

   A   lytic.
   B   benign.
   C   saprophytic.
   D   lysogenic.

Answer: D. lysogenic.
Lysogenic viruses remain dormant until something initiates it to break out of the host cell.

74. **Using a gram staining technique, it is observed that E. coli stains pink. It is therefore...**
   (Average Rigor) (Skill 19.2)

   A   gram positive.
   B   dead.
   C   gram negative.
   D   gram neutral.

Answer: C. gram negative
A gram positive bacterium absorbs the stain and appears purple under a microscope because of its cell wall made of peptidoglycan. A gram negative bacterium does not absorb the stain because of its more complex cell wall. These bacteria appear pink under a microscope.

75. **Segments of DNA can be transferred from the DNA of one organism to another through the use of which of the following?**
(Rigorous) (Skill 19.3)

    A   bacterial plasmids
    B   viruses
    C   chromosomal sharing
    D   single strand DNA

Answer: A. bacterial plasmids
Plasmids can transfer themselves (and therefore their genetic information) by a process called conjugation. This requires cell-to-cell contact.

76. **Protists are classified into major groups according to…**
(Average Rigor) (Skill 19.4)

    A   their method of obtaining nutrition.
    B   reproduction.
    C   metabolism.
    D   their form and function.

Answer: D. their form and function.
The chaotic status of names and concepts of protists reflects their great diversity in form, function, and life style. The protists are often grouped as algae (plant-like), protozoa (animal-like), or fungus-like based on the similarity of their lifestyle and characteristics of these more defined groups.

77. **Laboratory researchers have classified fungi as distinct from plants because the cell walls of fungi _____ .**
(Rigorous) (Skill 19.5)

    A   contain chitin.
    B   contain lignin.
    C   contain lipopolysaccharides.
    D   contain cellulose.

Answer: A. contain chitin.
All of the possible answers are compounds found in various cell wells. Cellulose is found in all plant cell walls, while lignin is found only in vascular plants' cell walls. Lipopolysaccharides are found in the cell wall of gram negative bacteria. Chitin is the only coumpound unique to fungal cell walls.

78.

**Identify the correct characteristiscs for the plant pictured above.
(Rigorous) (Skill 20.1)**

A   seeded, non-vascular
B   non-seeded, vascular
C   non-seeded, non-vascular
D   seeded, vascular

Answer: B. non-seeded, vascular
The picture above is of a fern, Division Pterophyta, which are spore bearing
vascular plants.

79.   **The process in which pollen grains are released from the anthers is
called:**
(Easy) (Skill 20.2)

A   pollination.
B   fertilization.
C   blooming.
D   dispersal.

Answer: A. pollination.
Pollen grains are released from the anthers during pollination and carried by
animals and the wind to land on the carpels.

80. **Water movement to the top of a twenty foot tree is most likely due to which principle?**
     (Average Rigor) (Skill 20.3)

     A   osmostic pressure
     B   xylem pressure
     C   capillarity
     D   transpiration

Answer: D. Transpiration
Xylem is the tissue that transports water upward. Transpiration is the force that pulls the water upwards. Transpiration is the evaporation of water from leaves.

81. **Which of the following is not a strategy used by a young cactus to survive in an arid envrioment?**
     (Rigorous) (Skill 20.4)

     A   Stem as the principle site of photosynthesis.
     B   A deep root system to get at sources of groundwater.
     C   CAM cycle photosynthesis.
     D   Spherical growth form.

Answer: B. A deep root system to get at sources of groundwater.
A waxy spherical stem as the site of photosynthesis is an adaption that limits water loss, and allows for maximum water storage. CAM cycle photosynthesis allows for the plant to open the stomata at night thus limiting possible water loss due to evaporation. Some cacti will develop a taproot, however only after they have reached a point where it is needed to stabilize the plant.

82.   **Characteristics of coelomates include:**

I. **no true digestive system**
II. **two germ layers**
III. **true fluid filled cavity**
IV. **three germ layers**
(Average Rigor) (Skill 21.1)

A   I
B   II and IV
C   IV
D   III and IV

Answer: D. III and IV
Coelomates are triploblastic animals (3 germ layers). They have a true fluid filled body cavity called a coelom.

83.   **Which of the following is a disadvantage of budding as a reproductive mechanism, as compared to sexual reproduction.** *(Rigorous) (Skill 21.2)*

A   limited number of offspring
B   inefficient
C   limited genetic diversity
D   expensive to the parent organism

Answer: C. limited genetic diversity
By being an identical copy of the parent the gentic diversity of the individuals in an area is very limited. Depending on the species, it is possible for an organism that is reproducing by budding to produce a much larger number of offspring than an organism reproducing by sexual reproduction. As for efficiency and expense to the parent organism, every cell produced to create a budded offspring is used in reproduction. In sexual reproduction an excess of cells are produced to create viable offspring, probably a more expensive technique.

84.   Phylum Mollusca provides a chance to examine both open and closed circulatory systems (with most species having an open system and the cephalopods having a closed system). Which of the following is not shared by open and closed cirulatory systems in molluscs?
       (Rigorous) (Skill 21.3)

       A   Hemocoel
       B   Plasma
       C   Vessels
       D   Heart

Answer: A. Hemocoel
A hemocoel is the blood filled cavity that is present in animals with an open circulatory system. Unlike some other types of open circulatory systems, the molluscs have three blood vessels, two to bring blood from the lungs and one to push blood into the hemocoel.

85.   After sea turtles are hatched on the beach, they start the journey to the ocean. This is due to:
       (Easy) (Skill 21.5)

       A   innate behavior.
       B   territoriality.
       C   the tide.
       D   learned behavior.

Answer: A. innate behavior.
Innate behavior is inborn or instinctual. The baby sea turtles do not learn this behavior from their mother. They immediately know to head towards the shore once they hatch.

86. **Which of the following answers is best described by the following conditions: not under conscious control and able to maintain a long term contraction.**
   (Rigorous) (Skill 22.1)

   A   Fast twitch sketetal muscle
   B   Slow twitch sketetal muscle
   C   Smooth muscle
   D   Cardiac Muscle

Answer: C. Smooth muscle
Smooth muscles are the only muscles that can maintain a long term (nearly permanent) contraction. Both cardiac and smooth muscle are not under conscious control. Although the control is often subconscious, both fast and slow twitch muscles are under conscious control, and cardic muscles can only contract for short periods of time.

87. **Which of the following compounds is not needed for skeletal muscle contraction to occur?**
   (Rigorous) (Skill 22.2)

   A   glucose
   B   sodium
   C   acetylcholine
   D   adenosine 5'-triphosphate

Answer: A. Glucose
Although glucose is necessary for the generation of ATP (adenosine 5'-triphosphate) it is not directly involved in muscle contractions. Acetylocholine is the neurotransmitter that intiates contraction. Sodium is an essential part of creating an action potential. Lastly, ATP is the energy source for the contraction.

88.   **Movement is possible by the action of muscles pulling on**
        (Average Rigor) (Skill 22.3)

        A   skin.
        B   bones.
        C   joints.
        D   ligaments.

Answer: B. bones.
The muscular system's function is for movement. Skeletal muscles are attached to bones and are responsible for their movement.

89.   **All of the following are found in the dermis layer of skin except:**
        (Easy) (Skill 22.4)

        A   sweat glands.
        B   keratin.
        C   hair follicles.
        D   blood vessels.

Answer: B. keratin.
Keratin is a water proofing protein found in the epidermis.

90.   **Capillaries come into contact with a very large surface area on both**
        the kidneys and lungs, especially in relation to the volume of these
        organs. Which of the following is not consistent with both organs
        and their contact with capillaries.
        (Rigorous) (Skill 23.1)

        A   Small specialized sections of each organ
        B   A large branching system of tubes within the organ
        C   A large source of blood that is quickly divided into capillaries
        D   A sack that contains a capillary network

Answer: D. A sack that contains a capillary network
The Bowmen's capsule of the kidneys can be described as a sack that contains a capillary network. The alveoli of the lungs are sacks, however the capillaries are on the outside of the alveoli.

91. **What is the principle driving mechanism of inhalation and exhalation?**
    (Easy) (Skill 23.2)

    A diaphragm
    B the muscles surrounding the ribcage
    C the lungs
    D the bronchial tubes

Answer: A. diaphragm
The thoracic cavity holds the lungs. The diaphragm muscle below the lungs is an adaptation that makes inhalation possible. As the volume of the thoracic cavity increases, the diaphragm muscle flattens out and inhalation occurs. The correct answer is (A).

92. **Which of the following parts of the kidney is responsible for the filtering of salt from the body?**
    (Average Rigor) (Skill 23.3)

    A proximal convoluted tubule
    B the loop of Henle
    C the distal tubule
    D Bowman's capsule

Answer: B. the loop of Henle
The Bowman's capsule contains the glomerulus, a tightly packed group of capillaries in the nephron. The glomerulus is under high pressure. Water, urea, salts, and other fluids leak out due to pressure into the Bowman's capsule. This fluid waste (filtrate) passes through the three regions of the nephron: the proximal convoluted tubule, the loop of Henle, and the distal tubule. In the proximal convoluted tubule, unwanted molecules are secreted into the filtrate. In the loop of Henle, salt is actively pumped out of the tube and much water is lost due to the hyperosmosity of the inner part (medulla) of the kidney. As the fluid enters the distal tubule, more water is reabsorbed. The correct answer is (B)

93.   **Which of the following is not generally a symptom of nephritis?** *(Average Rigor) (Skill 23.4)*

    A   Hypertension
    B   Hematuria
    C   Edema
    D   Necrosis

Answer: D. Necrosis
Nephritis usually occurs in children. Symptoms include hypertension, decreased renal function, hematuria, and edema. Necrosis is simply the dying of flesh. Thus, the answer is (D).

94.   **In which of the following blood vessels is blood pressure greatest?** *(Easy) (Skill 24.2)*

    A   veins
    B   venules
    C   capillaries
    D   arteries

Answer: D. arteries
Arteries carry blood to such a large number of capillaries that the blood flow velocity actually decelerates as it enters the capillaries. Blood pressure is the hydrostatic force that blood exerts against the wall of a vessel. Blood pressure is greater in arteries. It is the force that conveys blood from the heart through the arteries and capillaries. The correct answer is (D).

95. Congestive heart failure (CHF) is a devastating disease which, depending on the stage in which it is diagnosed, can be fatal in days or even hours. Which of the following is the most likely treatment to be prescribed by doctors in the beginning stages?

I   Surgery
II   Chemotherapy
III   Diet changes
IV   Diuretics
(Rigorous) (Skill 24.3)

A   I, II
B   I, III
C   I, II, III, IV
D   III, IV

Answer: D. III, IV
Congestive heart failure (CHF) is a condition that can result from any structural or functional cardiac disorder that impairs the ability of the heart to fill with or pump a sufficient amount of blood through the body. Most often this is caused by a buildup of fluid around the heart. While surgery to draw off fluid or transplant a new heart may be necessary as a life saving measure in the later stages, initial treatment usually consists of dietary changes to eliminate those foods that might cause the patient to retain water, and diuretics to help lessen the amount of fluid in the body, relieving part of the pressure (the congestion). The answer, then is (D).

96. A school age boy had chicken pox as a baby. He will most likely not get this disease again because of
(Average Rigor) (Skill 24.4)

A   passive immunity.
B   vaccination.
C   antibiotics.
D   active immunity.

Answer: D. active immunity.
Active immunity develops after recovery from an infectious disease, such as the chicken pox, or after vaccination. Passive immunity may be passed from one individual to another (from mother to nursing child).

97.   In which of the following situations would you expect doctors to use immunosuppresants?
        (Rigorous) (Skill 24.5)

        A   a skin graft
        B   a blood transfusion
        C   an organ transplant
        D   severe allergic reactions

Answer: C. an organ transplant
The immune system attacks not only microbes, but also cells that are not native to the host. This is a problem with skin grafts, organ transplantations, and blood transfusions. Antibodies to foreign blood and tissue types already exist in the body. If blood is transfused that is not compatible with the host, these antibodies destroy the new blood cells. There is a similar, but separate, reaction when tissue and organs are transplanted. The major histocompatibility complex (MHC) is responsible for the rejection of tissue and organ transplants. This complex is unique to each person. Cytotoxic T cells recognize the MHC on the transplanted tissue or organ as foreign and destroy these tissues. Various drugs are needed to suppress the immune system so this does not happen.

98.   Which of the following substances in unlikely to cause negative consequences if over-ingested?
        (Rigorous) (Skill 25.1)

        A   essential fatty acids
        B   essential minerals
        C   essential water-insoluble vitamins
        D   essential water-soluble vitamins

Answer: D. essential water-soluble vitamins
Water-soluble vitams are often removed in the filitration process by the kidneys and thus very rarely build to a dangerous level. To many fatty acids can lead to obesity related health problems. Excess minerals can lead to a variety of different conditions, depending on the mineral injested. Water-insoluble vitamins are usually stored in fatty tissues and thus are not flushed from the body and can reach dangerous levels.

99.   **A muscular adaptation to move food through the digestive system is called:**
      (Easy) (Skill 25.2)

      A   peristalsis.
      B   passive transport.
      C   voluntary action.
      D   bulk transport.

Answer: A. peristalsis.
Peristalsis is a process of wave-like contractions. This process allows food to be carried down the pharynx and though the digestive tract.

100. **Fats are broken down by which substance?**
      (Average Rigor) (Skill 25.3)

      A   bile produced in the gall bladder
      B   lipase produced in the gall bladder
      C   glucagons produced in the liver
      D   bile produced in the liver

Answer: D. bile produced in the liver
The liver produces bile, which breaks down and emulsifies fatty acids.

101. **Which of the following is not a common gastrointestinal disease?**
      (Average Rigor) (Skill 25.4)

      A   ulcers
      B   malabsorption
      C   diverticulitis
      D   diabetic nephropathy

Answer: D. diabetic nephropathy
Diabetic nephropathy is a kidney disease, and it is one of the primary contributors to the need for dialysis treatments in the Western Hemisphere. Ulcers are a well known gastrointestinal disease. Malabsorption is a general name for any condition that prevents the body from properly intaking one or more compounds found in food (e.g. lactose intolerance). Diverticulitis is the inflamation of one of the pouches on the outside of the colon.

102. **The role of neurotransmitters in nerve action is:**
     (Average Rigor) (Skill 26.1)

     A   to turn off the sodium pump.
     B   to turn off the calcium pump.
     C   to send impulses to neurons.
     D   to send impulses to the body.

Answer: A. to turn off the sodium pump.
The neurotransmitters turn off the sodium pump, which results in depolarization of the membrane.

103. **Hormones are essential for the regulation of reproduction. What organ is responsible for the release of hormones for sexual maturity?**
     (Average Rigor) (Skill 26.2)

     A   pituitary gland
     B   hypothalamus
     C   pancreas
     D   thyroid gland

Answer: B. Hypothalamus
The hypothalamus begins secreting hormones that help mature the reproductive system and stimulate development of secondary sex characteristics.

104. **Organophosphates, which are used as insecticides, herbcides, and nerve gas, inhibit the breakdown of acetylcholine. This serves as a rather extreme example of which type of drug's action?**
     (Rigorous) (Skill 26.3)

     A   Psychedelic drugs
     B   Stimulants
     C   Depressants
     D   Narcotics

Answer: B. Stimulants
By preventing the breakdown of acetylcholine, the organophosphate prevents the deactivation of the nerves that acetylcholine stimulates.

105. **In the case of someone experiencing unexplained changes in body temperature, sudden changes in level of hunger, and a change in circadian rythyms. Which of the following structures is most likely to be the cause of these problems?**
(Rigorous) (Skill 26.5)

    A   hypothalamus
    B   central nervous system
    C   pineal gland
    D   basal ganglia

Answer: A. Hypothalamus
The pineal gland releases melatonin, which has been linked to sleep/wake patterns. The basal ganglia and central nervous system are structures regulating nerve impulses. Only the hypothalamus is for regulating all of these conditions.

106. **Which of the following has no relation to female sexual maturity?**
(Rigorous) (Skill 27.1)

    A   thyroxine
    B   estrogen
    C   testosterone
    D   luteinizing hormone

Answer: A. thyroxine
Thyroxine is the hormone most often associated with the regulation of the body's metabolism, and is not related to the maturation process. Luteinizng hormone stimulates cells in both the testes and ovaries. Estrogen is the hormone most frequently associated with female development. Studies have indicated that levels of testosterone increase as a female heads through puberty until she reaches her sexual peak and then begin to drop off.

107. **Fertilization in humans usually occurs in the:**
(Easy) (Skill 27.2)

A   cervix.
B   ovary.
C   fallopian tubes.
D   vagina.

Answer: C. fallopian tubes.
Fertilization of the egg by the sperm normally occurs in the fallopian tube. The fertilized egg is then implanted on the uterine lining for development.

108. **There are many dangers inherent in drug and alcohol abuse, not the least of which is that such abuse can harm a developing fetus. Which of the following might we be surprised to see in the child of an addicted mother?**
(Average Rigor) (Skill 27.3)

A   cerebral infarction
B   micropthalmia
C   microcephaly
D   Klinefelter's Syndrome

Answer: D. Klinefelter's Syndrome
Klinefelter's Syndrome is a genetic diease that is the result of non-disjunction during meiosis. All three of the other conditions are often caused by drug and/or alcohol abuse.

109. **Which of the following best explains the fact that environmental factors are more often the cause of male infertility than female infertility?**
(Average Rigor) (Skill 27.4)

    A   Women don't work in outdoor professions as frequently as men do, thus limiting their exposure to environmental toxins.

    B   Men don't bathe as often as women do, leaving toxins on their hair and skin.

    C   Men produce sperm throughout their lives, while women produce no new eggs after puberty.

    D   The external nature of the male reproductive organs places them more at risk of exposure to environmental toxins.

Answer: C. Men produce sperm throughout their lives, while women produce no new eggs after puberty.

In the modern world, men and women work in outdoor professions and bathe frequently in equal numbers, making answers (A) and (B) incorrect. Most environmental toxins are either ingested from contaminated food or water or breathed in, making answer (D) unrealistic. The correct answer is (C).

110. **All of the following are density dependent factors that affect a population except**
(Rigorous) (Skill 28.1)

    A   disease.
    B   drought.
    C   predation.
    D   migration.

Answer: B. drought.

Although drought would affect the amount of food available to a population (which creates a density dependent factor), the drought itself would happen regardless of the size of the population, and is thus density independent. Diseases and migration tend to occur more frequently in crowded populations. The amount of prey and the amount of predators will affect the level of the population.

111. **In the growth of a population, the increase is exponential until carrying capacity is reached. This is represented by a(n):**
(Average Rigor) (Skill 28.2)

    A   S curve.
    B   J curve.
    C   M curve.
    D   L curve.

Answer: A. S curve.
An exponentially growing population starts off with little change and then rapidly increases. The graphic representation of this growth curve has the appearance of a "J". However, as the carrying capacity of the exponentially growing population is reached, the growth rate begins to slow down and level off. The graphic representation of this growth curve has the appearance of an "S".

112. **A clownfish is protected by the sea anemone's tentacles. In turn, the anemone receives uneaten food from the clownfish. This is an example of**
(Easy) (Skill 28.3)

    A   mutualism.
    B   parasitism.
    C   commensalism.
    D   competition.

Answer: A. mutualism.
Neither the clownfish nor the anemone cause harmful effects towards one another (parasitism) and they both benefit from their relationship. Mutualism is when two species that occupy a similar space benefit from their relationship.

113.   Population densities are usually estimated using representative plots. If we use a 100 square mile portion of upstate New York as our representative plot, what disperal pattern would you expect? *(Rigorous) (Skill 28.4)*

   A   random
   B   staggered
   C   clumped
   D   uniform

Answer: C. clumped
Upstate New York is populated similarly to many other non-urban areas of our country, with small towns having formed and people living in or near those towns. Because of this, we would expect to see a clumped pattern, as the population would not be so accomodating as to spread themselves out evenly around the area, nor are there many cases of random dispersal of people. Staggered is not a commonly used dispersal pattern. The answer, therefore, is (C).

114.   If DDT were present in an ecosystem, which of the following organisms would have the highest concentration in its system? *(Average Rigor) (Skill 29.1)*

   A   grasshopper
   B   eagle
   C   frog
   D   crabgrass

Answer: B. Eagle
Chemicals and pesticides accumulate along the food chain. Tertiary consumers have more accumulated toxins than animals at the bottom of the food chain.

115. **Which trophic level has the highest ecological efficiency?**
(Average Rigor) (Skill 29.2)

    A   decomposers
    B   producers
    C   tertiary consumers
    D   secondary consumers

Answer: B. Producers
The amount of energy that is transferred between trophic levels is called the ecological efficiency. The visual of this is represented in a pyramid of productivity. The producers have the greatest amount of energy and are at the bottom of this pyramid.

116. **High humidity and temperature stability are present in which of the following biomes?**
(Easy) (Skill 29.3)

    A   taiga
    B   deciduous forest
    C   desert
    D   tropical rain forest

Answer: D. tropical rain forest
A tropical rain forest is located near the equator. Its temperature is at a constant 25°C and the humidity is high due to the rainfall that exceeds 200 cm per year.

117. **Which of the following terms does not describe a way that the human race has had a negative impact on the biosphere?**
(Rigorous) (Skill 29.4)

   A   Biological magnification
   B   Pollution
   C   Carrying capacity
   D   Simplifcation of the food web

Answer: C. Carrying Capacity
Most people recongize the harmful effects that pollution has caused, especially air pollution and the concept of Global Warming. Pollution and regular use of pesticides and herbicides introduce toxins to the food web, biological magnification relates to how the concetration of these toxins increases the farther you move away from the source so that animals at the top of the food chain, for example Bald Eagles, develop dangerous levels of toxins, and maybe responsible for declining birth rates in some species. Simplification of the food web has to do with a small variety of crops replacing large habitats, and thus shrinking or destroying some ecosystems. Carrying capacity is simply a term that relates to the amount of life that a certain habitat can sustain, it is a term independent of human action, so the answer is (C).

118. **Which term is not associated with the water cycle?**
(Easy) (Skill 30.1)

   A   precipitation
   B   transpiration
   C   fixation
   D   evaporation

Answer: C. Fixation
Water is recycled through the processes of evaporation and precipitation. Transpiration is the evaporation of water from leaves. Fixation is not associated with the water cycle.

119. **In the nitrogen cycle, decomposers are responsible for which process?**
(Rigorous) (Skill 30.2)

A   Nitrogen fixing
B   Nitrification
C   Ammonification
D   Assimilation

Answer: C. Ammonification
Nitrogen fixing and nitrification are primarily handled by bacteria (although there are inorganic processes that augment this somewhat). Assimilation is a process carried out by plants. Ammonification is the process that falls to decomposers, thus preparing nitrogen compounds for the bacteria.

120. **Which biogeochemical cycle plays the least part in photosynthesis or cellular respiration?**
(Rigorous) (Skill 30.3)

A   Hydrogen Cycle
B   Phosphorous Cycle
C   Sulphur Cycle
D   Nitrogen Cycle

Answer: C. Sulphur Cycle
Although sulfur is used in some proteins used in photosynthesis and respiration, only in some photosynthesizing bacteria (not cyanobacteria) does sulfur play a significant role. Hydrogen is readily passed through almost every stage of respiration and photosynthesis. Many of the energy storing moecules (including ATP) use phosphorous as a principle component. Nitrogen, in addition to being a significant component of the proteins that assist in these processes, is also a significant component of NADP.

121. **The concept that the rate of a given process is controlled by the scarcest factor in the process is known as _____?**
      (Average Rigor) (Skill 30.4)

      A   The Rate of Origination.
      B   The Law of the Minimum.
      C   The Law of Limitation
      D   The Law of Conservation

Answer: B. The Law of the Minimum
A limiting factor is the component of a biological process that determines how quickly or slowly the process proceeds. Photosynthesis is the main biological process determining the rate of ecosystem productivity, the rate at which an ecosystem creates biomass. Thus, in evaluating the productivity of an ecosystem, potential limiting factors are light intensity, gas concentrations, and mineral availability. The Law of the Minimum states that the required factor in a given process that is most scarce controls the rate of the process.

122. **The biggest problem that humans have created in our environment is:**
      (Average Rigor) (Skill 31.1)

      A   Nutrient depletion and excess.
      B   Global warming.
      C   Population overgrowth.
      D   Ozone depletion.

Answer: A. Nutrient depletion and excess.
Human activity affects parts of the nutrient cycles by removing nutrients from one part of the biosphere and adding them to another. This results in nutrient depletion in one area and nutrient excess in another. This affects water systems, crops, wildlife, and humans. The answer is (A).

123. **The three main concerns in nonrenewable resource management are conservation, environmental mitigation, and _____.**
(Rigorous) (Skill 31.2)

A   Preservation
B   Extraction
C   Allocation
D   Sustainability

Answer: C. Allocation
The main concerns in nonrenewable resource management are conservation, allocation, and environmental mitigation. Policy makers, corporations, and governments must determine how to use and distribute scare resources. Decision makers must balance the immediate demand for resources with the need for resources in the future. This determination is often the cause of conflict and disagreement. Finally, scientists attempt to minimize and mitigate the environmental damage caused by resource extraction. The answer is (C).

124. **Which of the following is/are reason/s to maintain biological diversity?**

I.   **Consumer product development.**
II.   **Stability of the environment.**
III.   **Habitability of our planet.**
IV.   **Cultural Diversity.**
(Rigorous) (Skill 31.3)

A   I and III
B   II and III
C   I, II, and III
D   I, II, III, and IV

Answer: D. I, II, III, and IV
Biological diversity is the extraordinary variety of living things and ecological communities interacting with each other throughout the world. Maintaining biological diversity is important for many reasons. First, we derive many consumer products from living organisms in nature. Second, the stability and habitability of the environment depends on the varied contributions of many different organisms. Finally, the cultural traditions of human populations depend on the diversity of the natural world. The answer is (D).

125. **Stewardship is the responsible management of resources. We must regulate our actions to do which of the following about environmental degredation?**
    (Average Rigor) (Skill 31.5)

    A   Prevent it.
    B   Reduce it.
    C   Mitigate it.
    D   All of the above.

Answer: D. All of the above.
Stewardship requires the regulation of human activity to prevent, reduce, and mitigate environmental degradation. An important aspect of stewardship is the preservation of resources and ecosystems for future generations of humans. Therefore, the answer is (D).

CPSIA information can be obtained
at www.ICGtesting.com
Printed in the USA
FFOW04n1406130514
5356FF